YOUR FUTURE IN TELEVISION CAREERS

YOUR FUTURE IN TELEVISION CAREERS

By
DAVID W. BERLYN

RICHARDS ROSEN PRESS, INC.
New York, New York 10010

ST. PHILIPS COLLEGE LIBRARY

Published in 1978 by Richards Rosen Press, Inc.
29 East 21st Street, New York, N.Y. 10010

Copyright 1978 by David W. Berlyn

All rights reserved. No part of this book may be reproduced in any form without written permission from the publisher, except by a reviewer.

First Edition.

Manufactured in the United States of America

Library of Congress Cataloging in Publication Data

Berlyn, David W
   Your future in television careers.

   Bibliography: p.
   1. Television broadcasting—Vocational guidance—United States. I. Title.
HE8700.8.B47   1978   384.55'4'023   77–13818
ISBN 0–8239–0404–0

About the Author

DAVID W. BERLYN'S experience in the communications industry ranges from journalism to advertising sales management. He first reported in the news field as a U.S. Army war correspondent in England during World War II. He was graduated from the Boston's University School of Journalism and worked as reporter for a New England newspaper. He became successively a radio news writer, a regional news service editor, and a writer-editor with *Broadcasting* magazine, all in Washington, D.C. He was senior editor with the magazine in New York and now is Eastern Sales Manager.

Berlyn has lectured on communications at several colleges in Westchester County in New York and was contributor of the articles Television and Radio Broadcasting for *Collier's Year Book* (1974 and 1975 editions). He resides in Tarrytown, N.Y., where he is active in community affairs.

Acknowledgments

To Mary, Steve, and Debbie. To Bert Briller of the Television Information Office, without whom there would not have been this work. To A. Frank Reel, a good friend and industry sage. To Ruth Rosen, a person with patience.

Contents

	Preface	ix
	Introduction	xiii
I.	The Lure of Television	3
II.	Programming	9
III.	News	15
IV.	Management	22
V.	Engineering	29
VI.	The New Television Technology	37
VII.	Sales	43
VIII.	Promotion	48
IX.	Unions	54
X.	Women in Television	60
XI.	The Dimensions of Television	67
XII.	For Those Who Manage: Dollars	74
XIII.	The Noncommercial Station	77
XIV.	Widening the Field of Vision	83
XV.	Attitudes and Attributes	87
XVI.	Preparing for Careers	93
XVII.	Taking It from the Top	100
	Bibliography	113

Preface

This volume opens the door for ambitious young men and women and gives them a glimpse of the possibilities inherent in a career in the television industry. It is rich in detail in its description of the various branches of the business and the types of employment that are available. But before you decide to plunge into broadcasting, you should understand a few fundamentals that are necessary to a full appreciation of what you are about to face.

During my quarter of a century in the television business, and especially during the last seven years when I was president of Metromedia Producers Corporation, a company engaged in the production and distribution of filmed and taped programs and series for television, I interviewed hundreds of eager job applicants. Most were fresh out of school, where they had taken one or more courses in communications or some other aspect of the electronics industry. I found that although they were prepared to enter the ranks of creative or production workers in an operational sense, they were not ready to face the practicalities of the business. In short, for the most part they viewed the challenge as that of producing the best in terms of their own taste, without regard for the limitations imposed by the laws of economics.

To be truly prepared to enter the television industry in any capacity, you must understand two fundamental facts. The first is that American commercial television is a business operated for profit; the second is that there is a physical limitation on the number of available broadcast channels.

With respect to the first point, one should keep in mind that television executives are answerable to their stockholders, not to the critics. Commercial television in the United States is supported by advertisers, who use the medium in order to sell their products. The amount of money they pay depends upon the size and buying power of the audience, usually figured at so many dollars per thousand viewers, with additional emphasis on the demographic breakdown of those viewers. If, for example, the product is a food or detergent or cosmetic or medicine, how many women between the ages of 18 and 49 will view the ad?

The fact that the business must make a profit dictates the scramble for ratings. It explains why more money can be spent on production for prime-time than for daytime or late-night programming, on VHF (very high frequency) rather than UHF (ultra high frequency) stations, and on programs that are surrounded by rating "hits" on the same station or network on the same day, than can be spent on the struggling competition. And it explains why some "good" shows die while some "bad" ones live and why, consequently, our ideas of what is "good" and "bad" may themselves have to undergo reexamination.

The limitation of available channels is an even more important factor, partly because the concept is so foreign to the average American mind that finds it difficult to accept any limits to our genius for technical "progress." The salient fact is that there are only twelve useful VHF television channels, and that to avoid interference no two stations can share the same channel unless they are at least 200 miles apart. The result is that the great majority of U.S. cities are served by no more than three VHF stations. There is no way to get any more. New York and Los Angeles are unusual in that each has seven VHF stations. No other city has more than four, and only twelve have four VHFs. Market areas as populous as those centered by Philadelphia, Boston, Detroit, Cleveland, Cincinnati, Houston, Pittsburgh, Baltimore, Milwaukee, Kansas City, and Buffalo have three VHF stations. Dozens of other areas have only two. That is why we cannot have more than three networks if we accept the truism that a national television network must be geared to the VHF band.

Unfortunately we must accept it. UHF offers the prospect of seventy more channels, but when the Federal Communications Commission first allocated channels for this new medium at the end of World War II, UHF had not yet been perfected and the FCC didn't want to wait. By the time the FCC got around to providing for UHF, there were 17 million television sets in the country and none of them could receive a UHF signal. The three networks and the major stations had, meanwhile, spreadeagled the field on a VHF basis, and as a result the UHF operators—except in a few isolated instances—never got off the ground. This means that the UHF stations do not command as large an audience as do the VHF broadcasters; therefore they cannot get as much money from advertisers, and like a rolling snowball this dictates that they cannot spend as much on programming, which further diminishes their chance to increase the audience.

Thus, because of channel limitation we are led to the most significant characteristic of the television industry. There cannot be more than three networks, and, as a result, they are insulated against chances of further competition from without. This makes them the most powerful as well as the most important elements in this complicated business. The networks compete among themselves for the largest share of what is essentially the same audience, which inevitably has the effect of narrowing the medium's appeal and causing a discernible sameness in approach to programming. Eager and creative young people who come into the industry without a realization of this situation are due for massive disillusionment.

Knowledge does not guarantee success. But if you go into this business knowing that it *is* a business—for profit—and if you understand the implications of channel limitation, you are at least starting with your eyes open and an appreciation of the rules of the game.

A. Frank Reel

Introduction

What kind of jobs are available in television? What are your chances of finding one?

Performers, announcers, reporters, weathercasters are in evidence every time you turn on the television and flick the dial. It seems simple, but the television program is actually produced by a very complex business organization: the television station. It requires every skill that most American businesses utilize, plus a few more that are highly specialized.

A full business and managerial and creative staff is, of course, a necessity. This includes salespeople, researchers to support the program department and sales force, advertising and promotion specialists, news editors, and writers.

Camera, lighting, and control specialists; film and videotape editors; and technical and engineering personnel must operate and maintain intricate equipment. Specialized craftsmen, artists, set designers, lighting technicians, and electricians are required, not to mention filing and scheduling clerks and receptionists.

All play their role. Many stations and broadcast groups augment their staffs with data-processing personnel and computer programmers and services. Television offers opportunities for men and women in every one of these departments and specialties.

Obviously not all of the jobs in television are glamorous, but many offer full and satisfying careers. Television traditionally has been proud of the people who represent it, people who perform their duties conscientiously and well, often under great pressure.

Job prospects for the future are bright in the long run. This is especially so in the continually expanding areas of news, videotape production, and electronic data processing.

Television is a relatively young business, yet one with a tradition, dating back to its inception, of welcoming young people into its ranks. It is also an equal-opportunity employer, among the leaders in the hiring of women and members of minority groups.

It is important that you, as a potential job applicant, always think in terms of what you can offer in the way of specific skills to a network, to a station, or to a related enterprise. Merely presenting yourself as "wanting to get into television" is almost always futile; at best, it is wasteful in time and energy. Most stations simply cannot afford the space, facilities, and time to serve as a training school.

Although practical experience is, of course, a great teacher, training for a television career can begin at home or at school—and the earlier the better.

Education can guide you into developing your innate skills and talents and into recognizing your general areas of aptitude. Because television is an industry of communications, the ability to express yourself orally and with the written word is invaluable. Beyond that, you should attempt to define for yourself the area of television that most interests you and then hone your basic skills in that direction. For example, a ratings researcher must have a foundation in mathematics. A newswriter should know local and state history with an accent on politics and have skills in typing and grammar.

A high level in such basic skills is a sign of your sincerity and determination. It gives any applicant an edge in this highly competitive field.

YOUR FUTURE IN TELEVISION CAREERS

CHAPTER I

The Lure of Television

Do you imagine yourself on television? Perhaps you see yourself interviewing a prominent political figure or a world-renowned celebrity. Or acting in a taut drama, or performing in a rollicking musical.

You may picture yourself behind the camera making the star look good. You may see yourself in the role of writer who dreams up stories people will want to watch. Or the director who brings stars and scripts, lights and cameras together to make great entertainment.

These are just a few, and the more glamorous, of the many opportunities that may attract you to television. Many others offer exciting possibilities for rewarding work in television. This book will tell you a lot about the field to help you decide whether it's one in which you would like to fashion a career.

As a career field television is big and growing bigger. That means there will be job openings and opportunity for change and for advancement.

Television is diverse. You can find positions in broadcasting that require the ability to talk persuasively and others that don't require you to open your mouth as long as you can write interestingly. You can find positions in a studio where it won't matter how "far out" your style of dress may be and other areas, as in perhaps sales, where your clothes and how you wear them do matter.

4 YOUR FUTURE IN TELEVISION CAREERS

Television has the happy potential of welcoming people of nearly every skill and personality. It wants "numbers people" and "idea people," "desk people" and "outside people," "nine-to-fivers" and "work-round-the-clockers." And what makes television fun, too, is that it brings different types together. Sometimes it makes the nine-to-fiver work until midnight. It may make the idea person learn and

NBC PHOTO

In the control room all the elements of television come together to present a program of entertainment or information. Various camera shots of a program in action are portrayed on monitors to directors and producers. The choice of picture and camera angle is made here as the program is in progress.

like the discipline of working out numbers. Often TV calls upon the people working in it and with it to grow.

Growth of the Medium

Growth is probably the most important aspect of a television career. Television is in fact a growing industry, as much a business as it is an art or communications medium. Television is in flux, changing even as technology and society's needs are changing.

Television's growth has been rapid. In 1949 there were but 50 stations and 1,000,000 television sets. Currently there are more than 900 stations and more than 68 million television households, many with more than one set!

Because 98 percent of all homes have at least one television set and over 40 percent have two or more sets, additional growth must come in different areas. There are thousands of cable systems, and these are increasing in number and size. Television is developing new videotape usage and additional closed-circuit systems in which programs shown are not broadcast but relayed for specific purposes. Television's role in the schools and in business has been on the increase. This continued growth of the medium should create new opportunities for varied careers.

Salary Considerations

It is tempting to generalize about pay in television. But one could say for certain that people in television are paid somewhat higher salaries than are others employed in similar work in other fields. In part, this is because of the large audiences—and hence increased revenues—that television programs attract. It is logical that a person who appears in a television show entertaining tens of millions of people would earn substantially more than the performer in a theater who plays to an audience of perhaps two thousand.

Naturally salary scales vary by craft, by size of station and of the community it serves, and according to whether the station operates commercially or educationally, or is publicly financed with some advertiser "underwriting."

Those who engage in performing arts in television are affected directly by pay scales tied to a star's magnetism and hence his or her worth as a "talent." A top-name comedian or show host or hostess, even a network-television news personality, may be paid $500,000 or more a year.

As to the geography of television, generally, the 900 television stations in the United States, including Alaska, Hawaii, and Puerto

Rico, are where the people are. They are concentrated in the major cities; stations are fewer and farther between in the less populous states. It may be possible to tailor your personal geography and interests accordingly. If skiing is your thing, you could aim for a station post in Maine or Colorado, but if it's surfing that turns you on, perhaps stations in southern California should be your target area. There is, then, a wide though hardly limitless choice of locales from which you can choose those at least that have a first appeal. But you probably will not land the dream job in the dream town in the dream state or region on your first or subsequent attempts.

Advancement

What are the opportunities for advancement? Quite good. The television networks and many of the larger stations by policy recognize the skills and experience of employees. Performance on the job is important. The writer who can edit a script better than the average employee in his department will be moved up to an editing position; the staff member who can more often select those programs that will draw the largest audiences will find increased responsibility in programming and, of course, will earn a larger paycheck.

Most people in television work in relatively attractive surroundings and with little on-the-job risk. There are exceptions, of course, such as news reporters who cover fires, riots, and disasters as part of their jobs.

Many opportunities in TV also have the potential of providing you with a unique sense of belonging, of informing, and even of educating others, of knowing what is going on and being right where the action is. It may be that in the TV job you land, you'll have the enjoyment of seeing your work on the air and observing the influence it has on others. Television can be a productive outlet for various artistic talents, latent or realized.

As a volatile, fish-bowl industry, television tends to be honest and aboveboard. This is so despite the substantial profits to be made and the overwhelming opportunity to wield power and influence over the opinions and beliefs of the people served. The reason

The Lure of Television 7

NBC PHOTO

Television is people, action, and events—all three of which are very much in evidence at a 1976 national political convention as captured on the monitor.

is in part traceable to the American tradition of freedom of communications and freedom of the press balanced against the fact that TV is an industry highly regulated by the government. Stations and networks thus are subject to intensive investigation, and stations must obtain regular renewal of their licenses by the Federal Communications Commission. It could be self-defeating, if not self-destructive, to be other than candid and ethical in technical procedures and in business practices.

The many facets of television and its closely related fields have a special appeal for people who seek challenge. As a result, television is peopled with a good number of young, stimulating, and friendly persons who have varied interests and, as a consequence, tend to make television generally "a nice place to work."

Let's look now at some specific occupations that a television career offers. In general, they can be divided into these categories: (1) programming; (2) news; (3) management; (4) engineering; and (5) sales. We'll look at each in order.

CHAPTER II

Programming

If little girls are made of sugar and spice, then programming is what television is made of. The program is the ultimate product, as well as the showcase, of a television station, a network, or a television film company. And programming is made with, by, and usually *for,* money.

Maybe making money is not all that important to you. In a television career, it's likely that other factors such as prestige, satisfaction, or ambition, are the attractions. That's understandable.

The fact remains, however, that from the viewpoint of compensation the programming responsibility in television may be the most rewarding. (It's tops, too, in pay scale when compared with other businesses.) Key programming executives at a network are known to earn $200,000 or more a year. Although it is unlikely that there will be more than a few such highly-paid executives at the average network, there is the awesome fact that this level of pay exists at all. The TV networks, moreover, have another dozen or so ranking programming executives whose salaries start well up in the $30,000 range, which is also the salary level for heads of programming at major television stations—and in this era of escalating salaries, the levels may already be substantially higher!

In programming and production at the networks, job "specs" are

CBS PHOTO

It takes talent, cameras, lights, studio, and professionalism among other ingredients to produce a high-quality or mass-audience program in television. The scene above, on the set of the Captain Kangaroo *program (with the star at right), is indicative of the staging and equipment required.*

quite precise and in the main accurately indicate the area of responsibility and the duties. These are discussed in some detail later in this chapter.

Station Program Director

The station, however, is the place to which most young people turn for their first television opportunity. Although stations usually program but a small fraction of their schedules with in-studio productions, the knowledge of programming and program production is necessary and often critical to the station's operation.

The program director sits in the catbird seat at a station's program department. He is responsible for the station's product, whether it be entertainment, news, or public affairs shows produced by the station itself or by subsidiary or adjunct enterprises; network shows carried optionally by the station; or shows produced independent of the station or of a network and "sold" to the station by an outside company (television syndicator). In each situation, the guiding rule is that the station acts by choice—and often the program director makes that choice, with the station's checkbook in hand.

The program director works with the station manager and sales manager in determining and administering the station's programming policy. He shapes the schedule, develops new shows and improves old ones, and participates in the decisions involved in buying programs that are supplied by outside sources, such as the film-syndication companies. The director supervises his staff in matters of work assignments, budgetary considerations, and production of shows.

Though stations program film shows, they are required to originate some local shows including news, documentaries, community programs, panel shows, and the like.

There is no regular road to becoming a station's program manager. Logic would dictate the choice of a college that offers specialized curricula in radio and TV and some practical experience in TV operations. The future programmer would be well advised to concentrate on acting and to emphasize writing, announcing, and the analysis of program scripts and commercial scripts.

Other Programming Staffers

The production manager in the programming department oversees the many details involved in producing a station's programming needs in a broadcast day. He determines the requirements of personnel, space, and equipment, supervises studio activities, and coordinates the services of various departments such as scenic design, music, and costumes.

The producer-director plans and supervises the entire production of a program or series, or of commercials, that are produced at the station. This responsibility includes the selection of performers, the general planning of sets, lights, and properties, and the determination of the sequences of camera shots. The producer-director is the focal point of production. The person in this job must be both an ad-

12   YOUR FUTURE IN TELEVISION CAREERS

ministrator and a creative artist who coordinates the various program elements such as film or tape, scripts, and music.

The prop man, makeup artist, and costumer are generally hired on a free-lance basis and are not part of the station's staff. (Also, a scenic designer, if employed, plans and supervises the construction of scenery and the painting of backdrops.)

NBC PHOTO

In a specially equipped studio, news announcers (left to right) Catherine Mackin, John Chancellor, David Brinkley, and Tom Brokaw co-anchor coverage of the national elections in 1976. The electronically controlled map in the background and vote-counting backboards are among the visual devices used to cover fast-breaking stories.

Network Level Jobs

In a description of job opportunities on the network level, the American Broadcasting Company prepared the following:

- Continuity acceptance. Responsible for clearance of all broadcast material (scripts, commercials, films, electrical transcriptions or recordings, tapes) for adherence to standards of good taste maintained by the network, the industry in general, and the regulatory agencies of interest.

Reader-editor. One reviews and clears program scripts, promotion copy, tapes, and ETS (electrical transcriptions) to assure conformance to policies, codes, and regulatory agencies' requirements.

Reader-editor. Another registers scripts, searches such material as books, plays, short stories, poetry and articles, program ideas, and formats to protect the rights of the author and the company.

Clearance editor. Reviews and clears commercial copy and investigates the acceptability of new advertisers, their products, and any premiums or offers.

Literary rights clerk. Searches titles of programs for prior use to insure originality and to avoid infringement of the right of privacy.

Program development. This department at the network is responsible for obtaining new properties (live and filmed shows) and developing them into shows. These potential programs are submitted by agents and packagers. The packagers are companies with the facilities to take an idea or format and develop it into a complete live or filmed show. Shows are screened by the program department, and arrangements are made to buy the property if it is of decided value.

Executive producer. Handles new property submissions from packagers and agents. They may amount only to a new idea and format for a program or to a live or filmed show. The producer prepares them for screening, supervises actual production of new shows that are not package-produced, maintains daily contact with the producers while working out changes in the format or a rescheduling of talent. He or she also obtains new film properties, prepares or supervises the preparation of pilots and screens them, and calls upon advertising agencies—along with network salespersons—to help with the selling of the newly acquired properties.

(Some explanation: A pilot refers to an idea for a program series that is produced on film or tape and shown as a full segment for advertisers and their agents to see and consider. Usually the pilot show is more expensive to produce than the average program in a series. The helping with the selling of newly acquired properties refers to advertiser interest in the program in terms of potential ratings, the show's possible image, and quality of production, among other things.)

> The program service manager. Supervises production of commercials on live programs and helps in their creation with advertising agencies, arranges and controls commercial integration within programs, coordinates agency services for the network and handles client relations with local stations on the programming of the client's shows.
>
> The business manager in the network's program development department is concerned principally with the buying and producing of shows. The manager is familiar with contracts; actual or estimated show costs; payments to performers, directors, and producers; and preparation of the operating budget; he also maintains liaison with business affairs, cost control, etc.

Still other specialties in programming include the commercial schedule supervisor, the department of production services, the business manager in production, the unit for studio services, the manager of the unit, the supervisor, the department of operations, its various managers and supervisors, among others.

CHAPTER III

News

Television news may appeal to you as a career in the field of communications. Then again, you may be turned off by the long, often grueling, hours required. The pressure to keep pounding out copy on a typewriter in the clatter of a newsroom while nearby a handsome young adult is on camera in a studio reading your news copy in mellifluous tones may not be your cup of tea.

Television news analyst Harry Reasoner some years ago wrote about young people and their choice of a news career in broadcasting. He advised in part:

> . . . Make sure you like it and can live it. Make sure you not only like news, but like radio and television news all the time. Because no other field of news is so demanding, so irritating, so confusing—and, when everything goes right, so beautifully satisfying.

Team journalism referred to by Reasoner in this particular essay involved in addition to a reporter, the cameraman, "who is increasingly a photo-journalist himself"; the producer-director, the film editor, the news editor, a writer, and in the case of live remote broadcasts, "a whole crew of skilled technicians." Make sure, he said, that "you can be part of a team. You can either learn to work together, with respect for specialized talents, or you flop."

Several developments in the broadcast field have radically influenced news in the manner of its presentation as well as the role of TV as a news medium. The emergence of new, lightweight videotape and microwave equipment and of easily transported cameras has created a demand for new skills in television reporting and editing. Cameras have been technologically reduced in size from the bulky, heavy, unwieldy giants to hand-held, less conspicuous (and more mobile) versions, with portable power supplies slung over the back of a reporter or technician. Already this equipment, plus miniature tape recorders, has helped expand the scope of on-air news reporting. Reporters have been afforded extra mobility, and the area of stories and events accessible for coverage has been expanded.

Opportunities in News

Where do you look for opportunities? For one thing, stations now offer internships, thus opening the way to on-the-job training. For many young people this could hold particular glamour and excitement as well as a chance to become part of a vibrant, throbbing young industry.

The outlook is affected, too, by the staggering output of news and sports television coverage. Television stations usually originate and produce their own news programs and documentaries plus specials, special features, live-action coverage, and the like. Despite the drain on manpower and the expense in facilities, the networks cover special events—such as presidential inaugurations, news conferences or state funerals—almost as a public service as well as for prestige.

Considerable talent, work, and skill and the performance of many jobs and assignments go into TV news preparation. There's more than meets the eye of the viewer that is trained on the home screen. An army of men and women, a battery of sophisticated studio equipment, of systems and interconnected relays, are employed to bring to TV-life the statesmanship of a White House address; the color of a Peking conference; the dash and style of Olympic Games held abroad; the familiarity of evening news with Walter Cronkite or

John Chancellor, and, of course, the many scenes of Americana—the weather forecaster predicting sunny skies; the local kennel show; a county fair in the capital city.

A good many viewers must imagine that news reporting is easy, glamorous, an opportunity to meet the biggest names, attend parties, and so forth. For the person toiling diligently in news, coverage and production appear to them to be the climactic point after numberless tiring hours, perhaps days, of scripts, takes, scenes and situations, an apparent throwing together of many talents and sequences.

A Day in News Production

The typical news day at an average television station follows a pattern. People such as the news director, who is a key to any TV-news operation, have jobs to do but their performance in even the most ordinary, routine duties may directly affect the station's output and quality of news programming.

At the station, the news director pulls things together. First thing each morning, he meets with editors and reporters, hands out assignments, and proceeds with the usual mound of paperwork. When staff people are not out during the day covering specific events, they may be assigned to work together on a news story for a future special report or a half-hour documentary.

The director has made his assignments by 9 A.M. or perhaps by 10. The reporters and camera crews are out, expected to have film back at the studio and ready to be put on the air for the evening broadcasts. The director's paperwork includes the makeup of work schedules, answering mail, and the usual reports to management on all aspects of the news operation. The day, and responsibilities, have just begun.

Television news *is* different. It is a news medium related to the newspaper and the magazine, but it is quite apart from print journalism in technique and has its own expertise.

Underlying Harry Reasoner's analysis was his belief that news work can be satisfying, and for many in the field the degree of satisfaction can be the most important factor, transcending the periods of drudgery, of routine reporting or writing or research.

When one attempts to place a judgment on any one facet of the television business, it becomes almost impossible to simplify. For example, the three "p's" in broadcasting influence those who work in the field: profits, programming, and public service. Television news operations are affected by all three; they should foster quality in programming, provide public service, and, wherever possible, turn a profit.

CBS PHOTO

It takes a team of news people to produce a news show. For the CBS evening news show, which features Walter Cronkite (left, background) as its anchorman, a host of reporters and writers work in the studio before the telecast.

The president of the National Broadcasting Company, Herbert G. Schlosser, in an interview published early in 1977, spoke of his wish for his company's improvement "in public service and the quality of programming, as well as improvement in profits," and added that, "actually, these values go hand in hand."

Television then is more than a news vehicle or the carrier of public-affairs programming. It is also *entertainment*. Moreover, it is a dynamic advertising medium and sales vehicle for goods and services. Yet a substantial portion of television content is devoted solely to news or news-oriented programming. With editorials and features added, the full impact of TV as a news medium is considerable.

News is said to be the fastest-growing part of the television industry. At most stations, it is a separate division that operates under a news director. Because of expansion, such as in the number of people who are in support roles for the growing number of on-camera reporters, job categories have multiplied. Consequently, the average TV news operation may have news writers, editors, producers, directors, film and videotape cameramen and editors, and lighting and audio and sound men, all in support roles for its on-air news reporters, anchorpersons, commentators, or editorialists.

Sensitivity to Change

Not only is the area that is devoted to news in a state of extraordinary expansion, but also it is sensitive to quick change. Current events can have immediate, profound effects on the medium and may well account for the way TV covers those news events. This may be seen in the way a President changes his methods of dealing with newspeople and arranging his news conferences. The coverage by a medium can be affected, for example, in a President's use of TV for international propaganda. Political conventions can be modified many times in a period of years in attempts to fit the events procedures to the discipline of the television clock (that is, for "peak audience viewing"). Television may be having a pronounced impact on the very fiber of the political process.

Thus, while events may effect changes in television's news makeup, the mere presence of cameras and supporting paraphernalia can create a sharp stimulus to people in the way they act or react. Often the awesome awareness of the TV reporter's entourage alters the behavior of participants of a telecast event.

As previously mentioned, newly designed small equipment—the minicamera and solid-state devices—is responsible for drastic changes in the news business. The emergence of lightweight videotape and microwave machines and of hand-held, midget power-pack cameras has created a demand for new skills in reporting and editing.

Inevitably, these changes expand the scope of electronic news operations.

The trends toward news-team assignments, expanded job categories, and increased turnover of news personnel are but a few in television news, which, if they persist into the 1980's, are bound to widen horizons further. They account, too, for wider opportunities

NBC PHOTO

News programming is responsible for a large share of overall television scheduling. Special events, such as national political conventions, increase the emphasis on news, as in the coverage of the 1976 political conventions by John Chancellor (left) and David Brinkley, anchormen at the National Broadcasting Company.

for minority groups in TV news. As the chapter on women in television points out, there are more women *reporters* on the home screen than at any time in TV history.

But the fact remains that those who are going to achieve success in television news must have a highly trained ability to write and to speak the English language with precision and clarity. They must have the desire and ability to involve themselves in local affairs so as continually to extend their station's area of news coverage while keeping pace with the constantly improving technical capacity of the medium.

Television News Personnel

News personnel come from a variety of sources. At one time the newspaper, the wire services, and occasionally radio were the most fertile feeders of news talent to television. Now the medium develops and trains its own people: graduates of journalism schools or holders of liberal arts degrees who show particular enthusiasm and an insatiable interest in people and places around them.

Television news people belong to a union, the American Federation of Television and Radio Artists (AFTRA), which has locals in most big cities. A rule of thumb at many TV stations' news operations is that job applicants be either well educated or have experience —and, of course, preferably both.

If the applicant has wide experience, that person may get a job without a full formal education. If one has no experience, it's a must to have a good education. Recommended is a sound liberal arts education. Although some technical jobs in broadcasting may not require such an education, in news it's almost a must.

It's best to apply for a job at the writer's level. As a writer, you can obtain a sound base, and it is more likely that a writer will be assigned to news coverage in the field.

Only a few years ago, union average pay for news people was $165 a week, and after three years $305 a week. Those with much experience received $100 or more over union scale. News announcers earn the big money in that area: from $1,500 and $2,000 a week and up. In the larger cities, the pay range for television news writers is higher, and as of 1977 it was being increased from a $400-a-week base to $500 a week after four years in such areas as New York, Chicago, and Los Angeles. (Under a new contract with the Writers Guild of America, news writers paid a minimum $400 under an old contract were to receive $430 a week in the first year of a new contract, with the weekly salary increasing until it reached $500 in the fourth year or in the 1980's.) On-air people, accordingly, also were negotiating upward, through AFTRA.

CHAPTER IV

Management

Broadcasting is big business. It is one of the highest paid of all U.S. industries. Its personnel roll is extensive. The major networks in broadcasting employ thousands. About one job in six in the industry is a network position. In television alone, the three national networks in the 1970's employed nearly 20,000 persons. This accounted for about three out of every ten staff employees in television.

Yet only a fraction of all jobs in broadcast *management* is in network operations. This is because broadcasting, while admittedly big, is in reality a composite of small enterprises. The typical number of employees at a TV station ranges from fifty to sixty. Since TV stations in the United States number in the hundreds, with each station individually operated and managed, it is obvious that a large segment of the industry provides positions in the management area. This localized makeup of TV stations means there is usually room for the bright young person to advance within a department into management ranks, often into a top-level job.

I find in my visits to stations around the country that just as most have a common outside appearance or location, the typical station has an internal table of organization that is also common to the entire industry. Usually this table starts at the top with the general manager or station manager.

The general manager runs the station. As a seasoned broadcaster, usually with sales, programming, production, or engineering experience, he evaluates his employees' recommendations and makes the final decisions on everything from station policy and programming to the purchase of new equipment. Reporting directly to him may be the controller, who oversees departmental costs, and the business manager, who handles the station's transactions. (This department operates much as in any other business, the only difference being that revenues are derived from the sale of advertising time rather than from a more tangible product or service.)

The table of organization includes managers of general sales, national sales, local sales, program, production, promotion, and business, a chief engineer, and a news director, or their equivalents. Generally these titles are used, although there may be variations. Occupations, too, are subject to variation, particularly at the larger stations where additional managerial responsibilities exist.

The General Manager

At the average station, the general manager is usually responsible to a board of directors, or if it is a large station group perhaps to a president or other chief executive, who in turn reports to a board. In any event, these people represent ownership, which expects the manager to run the station at a profit.

Because television is unique, the management function, though similar to those of other industries, has its own intricate character. Unlike counterparts in other industries, the manager of a television station is in command of a complex business that operates 365 days a year and, more often than not, 17 hours a day or longer.

Then, too, television's impact in the community cannot be minimized. It affects all segments of society, and for many people it occupies much of their waking hours.

The broadcast manager, ideally, is a person who can think clearly and see more than one side of an issue. A cultural background and thorough educational experience are invaluable for the general manager. This broadcaster must be able to maintain something resembling a harmonious relationship within the community. Outside of it, he usually seeks a rather similar and familiar association with the TV industry at large and with the United States government in particular. The latter is important. The broadcaster is a licensee of the Federal Communications Commission, and he must operate the station in "the public interest, convenience and necessity."

As one broadcaster, Milton D. Friedland, a vice president–general manager of Plains Television Stations, put it:

> It has been said that the station manager must be part educator and editor; psychologist, public relations [person], and sociologist; businessman, lawyer and engineer, and showman.

A manager should get to know and like people and should possess the quality of encouraging people to like and work with, and for, the manager. The manager must spend wisely and at all times be acutely conscious of costs. The person in command should understand and know sales, both in the actual sale of time to advertisers and in the broad spectrum in which there are sales principles.

Survey of Managers

This emphasis on sales appears in the background of a number of broadcast managers. Recently I independently researched the background of 25 management executives in the broadcast field. I found more than half (14) had at least some seasoning in sales. Several of the executives had come up through the ranks in sales, starting as an account executive selling time, and then moving up to local and subsequently general sales manager.

A study of backgrounds identified another characteristic: They made it to the top at a relatively young age. This bears out a traditional belief among those employed in television that their industry above others has always encouraged young people. Most broadcast managers, it is therefore believed, "arrive" in their chosen endeavor early in their careers.

In the study, I arbitrarily chose station manager or general manager as the upper ledge of the executive's climb in broadcasting. If these specific job titles did not appear, I took the closest equivalent.

The age of arrival thus was the age at which the individual was named station manager and not, say, general sales manager, which of course was recognition at an even younger age and at a time still earlier in a career.

Of the 25 executives, 12 could be judged managerial when they were between the ages of 35 and 44, and 10 were 34 years old or younger. At one end of the scale, 4 of the latter group of 10 arrived when they were between the ages of 25 and 28, and 3 were 33 or 34. Of the 3 who were 45 years old or over, only one, interestingly, was a woman and yet not a station manager. She gained her recognition as a manager of films at one of the nation's largest TV stations.

Another of those over 45 had come into broadcast management only after an executive career with the parent publishing company, and a third had spent his previous career in various submanagement jobs in radio operations.

In this comparison, I found it best also to take into account that most of the 25 persons studied were in jobs of managerial responsibility quite early in their careers. Research bears out the belief that broadcasting can be a young person's game.

Other than sales, the majority of executives surveyed had backgrounds in programming or production. Others were in announcing or engineering. A good many of them were unusual, and all had had a college education. They displayed, too, a combination of leadership, determination, and willingness for hard work.

One exceptional executive utilized his academic preparation and his talent quite early in life. When only 25 he took on management responsibilities at a New York City television station. He had been catapulted into the job only shortly after he had graduated from the Massachusetts Institute of Technology with training in business engineering administration.

This executive, Leavitt J. Pope, eventually was named president of WPIX in New York. He had started as an administrative assistant at the New York *Daily News,* where he had showed an interest in technical production and management, and he was assigned to construct new facilities and install presses for the newspaper. He had tinkered with ham radio as a youngster. It was a direction that when he was only 25 led this graduate of MIT's Sloan School of Industrial

Leavitt J. Pope, president of WPIX, and independent television station in New York City.

Management to leapfrog to a task force that put the newspaper-owned TV station on the air in 1949. He soon became WPIX's assistant manager, then operations manager.

According to the study, there is no pat way for the ambitious executive to enter TV management. A number of those surveyed were ushered in by the way of radio; fewer came in through TV's own channel. This ratio should reverse, however, as the television medium itself cultivates new management talent from within. (Significantly, Leavitt Pope was one of but two executives in the survey whose entrance to television was provided by publishing. Newspapers at one time served as a productive source for broadcast talent; they no longer do so.)

In broadcasting, where pay levels already are above the industrial average, management positions pay quite well. Management responsibility can be the most rewarding in money, with the exception only of the lofty salaries earned by star performers.

It is estimated that in the early 1970's one-fourth of all staff employees held managerial jobs, including the classifications of producer and director in addition to manager.

The station manager really is the "key blender," as one broadcast-industry expert observed, of the private interests of the owners with the interests of the public. As mentioned early in this chapter, the manager is responsible for the return in profits as well as the continuation of the broadcast license. It is important, too, for the manager and staff to establish a loyal and constant viewing audience for the station; otherwise the station may not make a profit and may eventually collapse.

The Program Manager

The basis for a station's economic health is programming. It is by that scheduling that the station may be ranked.

Enter the program manager. This manager's function is essential to the operation of the station and essential, too, in nurturing an effective and profitable management. The program manager is responsible for the broadcast day. This manager sees to it that local live shows are produced and directed and that all news elements are handled effectively.

Under the program manager's aegis, announcers are hired and fired, and the manager must see to it that announcers are properly scheduled. This programming executive buys film programs but only after considering audience tastes, show costs, and how best these shows fit into the station's total schedule.

Managers in sales and operations are departmental heads. The business manager primarily is responsible for providing financial advice and guidance to the station manager. He is the controller, and usually no money is committed or spent without his consent.

At the networks, the business and production manager supervises and maintains the budgetary responsibilities of the advertising/

promotion department. It is this executive who controls and checks the department's expenses on a day-to-day basis, supervises the acquisition of promotional material for network and local sales promotion, and supplies network and individual station requests for advertising and promotional materials.

Another programming executive on the network level is the business manager, who becomes familiar with program contracts—either the buying or producing of the show—and with the actual or estimated show costs. This manager handles payments to performers, directors, and producers; approves orders to make a pilot of a specific show, but not necessarily to buy it; becomes conversant with union contracts covering payments for production workers, and prepares and administers the programming department's operating budget. It is the business manager, in effect, who maintains a liaison with business affairs, cost control, personnel, operations, and accounting.

To pursue management in this field, you must be equipped with a good educational background. A liberal arts course with an emphasis on economics, marketing, and other facets of business administration would be useful in preparing yourself for such a career. If a course on broadcasting or advertising is available at your school, it may give you an edge.

If the opportunity is there at a station to start at the very bottom of the rung *but in sales,* perhaps as a trainee or a junior assistant in the department, consider it. Sales is an excellent avenue to travel in learning the broadcast business. It is also the traditional training ground for future managers.

CHAPTER V

Engineering

Top-echelon broadcast engineering jobs require formal education. For the person who strives toward that pinnacle, a sound academic background is essential. Preferably an applicant should have a degree in electrical engineering. For the better-paying positions or as a wise investment, moreover, hopefuls should become knowledgeable and proficient in computer technology. That is the thrust toward the future, and to some extent it's already the present.

If your goal is managerial responsibility, it is advisable to augment your bachelor's degree with a master's degree in business administration. By combining your engineering skill with business-administration excellence, and by supplementing both with know-how in computerization, you would be solidly equipped and ready to move ahead in TV engineering.

Even if your aim is modest—that is, targeted to the area where broadcast engineering jobs are more likely to develop—you'll need some education. A college degree is a sure plus.

It should be helpful to you, also, to know that people on the managerial level ought to be good managers first and crackerjack electrical engineers second. If you can be both, you'll be sought out. What's important, however, is this: the ability to lead and to relate to people could well spell the difference between the ordinary worker and the successful departmental manager or director.

If you don't have a degree, it is advisable to obtain at least some training in school. Secondary schools or training centers usually have facilities. Institutions of higher learning usually offer access to specialized training. Many colleges have broadcast stations on campus. Some schools offer radio-TV studies, some of them technical. If need be, take electronics as an outside interest or avocation.

Technical Challenges

Virtually every day brings the introduction of some new piece of electronic equipment into the exciting world of broadcast engineering. In the relatively young industry of TV, many consider the technical area the most promising and challenging. In order to gain the top technical rank in broadcasting, a First Class Radiotelephone operator's license is essential. To obtain such a "first ticket" requires considerable training, and the candidate must pass a rigorous written examination administered by the Federal Communications Commission. Local commission offices in the larger cities and the principal office in Washington can supply details about the training necessary for this important technical qualification.

That we now bounce TV signals off satellites as a matter of course should be a dramatic reminder that not only are we talking about a sophisticated business, but about an industry that is affected directly by space technology. Future broadcast technicians should come by an interest in this technology almost naturally. You should feel the urge to keep up with this continually expanding field. Space technology as a science should appeal to your imagination. It should present you with a challenge. It should appeal to you because of your career interest. If this area turns you off, then perhaps you ought to seek another career.

An additional point for you to keep in mind is that unions have a major role in the engineering departments and the technical operations of most stations and of all the networks. The chances are greater in this part of the broadcast business that you'll be required to join a union.

The union contract may contain language that affects job specifications (they may be spelled out in detail, in fact). Because of the union and the nature of the job, you may be required to serve a period of time as an apprentice before you become a regular operator. Your job advancement and salary would be affected accordingly.

When is the best time to look for a TV job in the technical field? If you are able to seek employment at a major network or at a large TV station while you are still in school, consider the possibility of applying as a vacation relief. Often this experience can lead to a permanent post. But apply in *January*. You must be an early bird, even earlier than the first pert robin of the Spring.

What the Jobs Are

The network is a giant broadcaster and employer. It is well staffed and covers all possible job functions. All broadcast operations, ranging from the average major TV station down to the tiny outlet, follow the networks' pattern. A description of job opportunities at the major TV network should be indicative of what is available on the local level, with the qualification that the numbers of people and of functions will vary with station size.

As a department, network engineering plans, designs, and either buys, builds, or installs plant facilities. Facilities include the network's real estate, offices, and technical operations. An extension of this includes mobile units, studios, and similar installations. A staff position normally requires a college degree, specifically a Bachelor of Science in electrical engineering. The staff engineer develops plans and prepares budget estimates and specifications of equipment to be purchased. He supervises the installation of video, video recording, film projection, and studio lighting systems.

Network operations provides manpower and facilities for programs, master control, projection room, and remotes. Job specs include technical director, studio-video and studio-audio technicians, and schedule clerk.

Responsibility for maintaining the technical quality of the program lies with the technical director. This person directs the activities of all technical personnel and maintains liaison with the program director. A studio-video technician runs TV cameras, handles microphones, and regulates the electronic quality of the television picture. A studio-audio technician similarly operates microphones in the studios and maintains the correct sound transmission level. A radio recording technician operates sound-recording equipment that reproduces live shows for delayed broadcast or for recorded programs. Similarly, the video recording technician handles kinescope and magnetic film recording equipment and optical sound recorders to reproduce live shows for delayed telecast and for distribution.

In another area, the maintenance section is responsible for the upkeep of electronic equipment to assure the highest possible performance standards. Applicants should have an electrical engineering degree and preferably some practical experience in the field.

The maintenance engineer is charged with equipment performance, which requires an ability to apply both theory and practical knowledge to problems met in keeping equipment up to par. The engineer must know how to correct a fault in the system and be able to make rapid analyses of circuit failures, diagnose the situation, and possess the know-how to make repairs.

Closely related are recording and film services, which vary in size, depending on the size of the broadcast company. At the network the department records and edits tapes, edits kinescopes after recording, edits features and prepares network films for broadcast, obtains film prints, provides scheduling and distribution to stations affiliated with the network, and controls the screening and quality of film. The major jobs in this department include a person in charge of editorial coordination, another for distribution, an order clerk, and the program editor.

Related to engineering is traffic and communications. On the network level, this function provides and arranges video and cable facilities by working with the American Telephone and Telegraph Company.

Engineering in Perspective

Simply stated, the engineer's job is to make it technically possible to get the program out of the studios, to the transmitter, and into the home.

Engineering 33

NBC PHOTO

At outdoor news events, the piggyback camera and power pack (carried here by a person accompanying the cameraman) have provided an extra dimension to television.

The job is essential in an industry that is at once exciting and challenging. No matter how routine the duties, how dull the experiences or mundane the surroundings, the broadcast technician is within grasp of opportunity to widen horizons. Though it is a tough, competitive business, broadcasting has its rewards, in satisfaction as well as in financial return.

In our lifetime, television is synonymous with innovation. The medium has been at its dazzling best when it has covered such major events as the moon landings; street riots; Presidential trips abroad; the debates between John F. Kennedy and Richard M. Nixon, and between Jimmy Carter and Gerald Ford; football Superbowl games, or the baseball World Series. Television feeds on events, on new and fresh talent, on sudden, perhaps electrifying developments on the world scene. That is the nature of the medium, and, accordingly, the engineering or technical know-how of the networks and TV stations has kept pace.

No longer is the source of programming confined to a stage in an enclosed studio. Any place, any setting in the world is now TV's stage. The First Lady tours the White House on camera, while millions watch and listen to her commentary. A movie to the technician is merely a roll of TV tape or film, but a national political convention he may experience as a mad but exhilarating scene in a kaleidoscope of newsmen, clutter, cameras, and hot coffee.

An assassin aims his weapon at a President while the camera dispassionately registers the infamous live action for millions to witness in shock and horror—not one of these events is "staged" or studio rehearsed.

The growth of the television medium need not be measured only in dollars and cents, nor necessarily by the number of rave reviews that programs receive from the critics. Another, more satisfying way, particularly for the student of the industry, is to take stock of the giant technical strides achieved by the medium.

In news, for example, developments have occurred with the onrush and buildup of a crashing ocean wave. Some were subtle and accepted by viewers in a ho-hum fashion with the attitude of "So what else is new?"

Changes that once were novel are now the technician's TV stock-in-trade. Graphics are more versatile and appealing to the eye; more sophisticated camera locations, equipment, and versatility add impact to picture quality as well as story content; split-screens and other optical techniques are now the ordinary practice in television program and production. And, on that world stage, the increased use of the hand-held news camera has permitted a wider range and flexibility in news coverage, and internationally it took less than ten years from the time of experimentation for news programming by satellite to become commonplace.

The quality of the television picture throughout all these changes and developments has improved dramatically. The viewer, who has become sophisticated and no longer so impressed by TV's amazing

NBC PHOTO

The control room at NBC during the network's coverage of the national political conventions in 1976. A news show is the product of a small army of trained technicians, writers, and newsmen.

progress, may not fully realize the social and political significance of satellite-relayed pictures from Peking that are as clear as the live transmissions emanating from his local TV station's news studio.

There has been a technical revolution also in television coverage of sports. The games themselves are played the same; the athletes have different names and numbers, but the game strategy hasn't changed all that much. What has been radically altered is the TV technique—the *engineering*. Most of these technical advances represented a development that worked and appealed to viewer and production expert alike and then added to the daily routine coverage. Yet the fan in the living room has been hardly aware of the changes and takes so many of them for granted: the camera's freeze-action; replay; on-field microphones; unorthodox camera locations and pickups.

Training for a career as a broadcast technician can offer high excitement. True, its effects may not be as breathtaking as those offered to teachers by the New Math when it was first introduced to the nation's classrooms—but it has its challenges. It could be most appealing to you if you like to make things work, if you take pride in what you can assemble or correct, despite the intricacy of a device.

Let's assume that in some technical jobs in the field, you don't have the opportunity to participate in or be part of television's glamour. There are other satisfactions. Job satisfaction for one. Reasonably good pay is another. There are the advantages of job security; because of the strong unions in the field and the nature of the work itself, it is almost certain you will be safe and insulated from the usual disruptions in employment in other areas. You will, for example, be less subject to the whim of the economy as compared to people in sales and programming, and not as susceptible to a change or new quirk in management.

The gamble and risk may be less in engineering, and the immediate rewards accordingly slightly less; but the opportunity is there for you to grow, to derive satisfaction and longevity on the job.

CHAPTER VI

The New Television Technology

Engineers in television are inevitably affected by science and by history.

A new television era began with the launching on July 10, 1962, of the 170-pound Telstar and with it, the introduction of satellite television. After the experimental communications satellite was rocketed into orbit, the news media reported that the feat of relaying the first transatlantic TV broadcast rivaled in significance the first telegraphed transmission by Samuel F. B. Morse more than a century before.

While television's trend in history has been to widen horizons and open man's mind, it would be well for the would-be broadcast engineer to keep one's own immediate goal within relatively modest reach.

To enter the field, you must first of all be ready to work hard and steadily at a job that at first may not seem to have much potential. You must be equipped; well prepared with a college degree, and if possible with some experience. The practice of just "catching on," if indeed it ever existed in the electronics business, had to become an anachronism as early as the 1950's and to be ignored by those in the field in the 1970's.

You must also be aware that even as television as an industry has experienced marked changes in recent years, so has engineering both as a science and as a profession. It was seemingly only "yesterday" that broadcast engineers were in demand. In the 1960's, for example, an acute shortage of engineers indicated that these highly specialized skills would be needed in abundance for years to come.

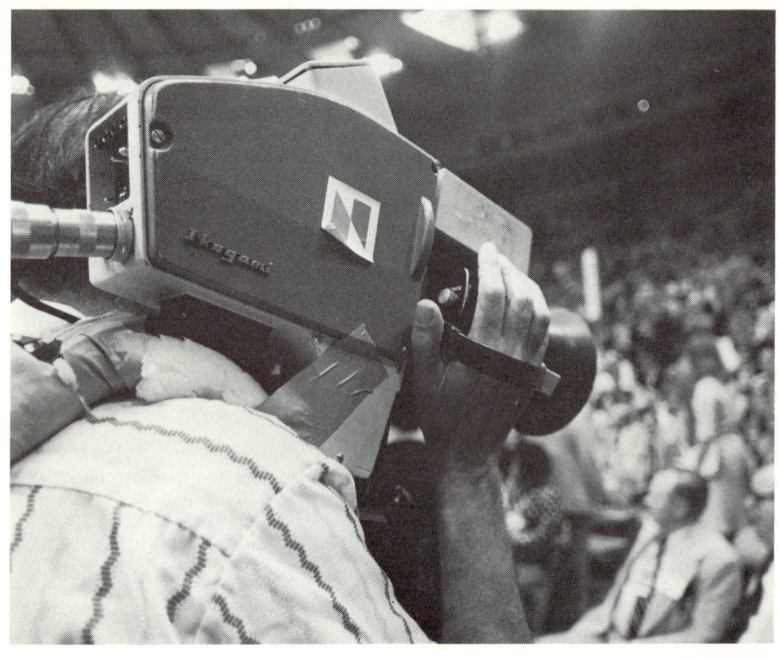

A closeup of a lightweight Japanese-made (Ikegami) camera that is hand-held and mobile.

But in the 1970's TV had to react to the manifestations of a stumbling economy in which growth patterns were not clearly defined or identified.

Julius Barnathan, president in charge of broadcast operations and engineering for the American Broadcasting Company, sees technology in broadcasting to be in a state of constant change. This has come about most emphatically in engineering design. Little imagination is needed, perhaps, to design a new electrical outlet in wiring a home or to figure out how to pour concrete. But in broadcasting the rapid breakthroughs in technological design have been mind-boggling.

The industry moved from tubes—used historically in radio and carried over naturally to television—to at first a tentative and then full emphasis on size and reliability as the transistor replaced the tube. From that time on, technological design in broadcasting accelerated.

Modules, integrated circuits, and logic board systems in which computer concepts play their part, followed. In a relatively brief time span, the industry spun away from the tube to microprocesses that make it easier to operate equipment. And, with digital design, the equipment was at once more reliable. Equipment was smaller and thus easier to handle, with fewer parts and fewer wires.

The Design Engineer

What has developed then is the emergence of a highly trained, highly skilled technician, new to and new in the broadcast field: the design engineer. It is this person who designs equipment or who understands in total what others already have designed in the new technology. (Suffice it to say that this person's credentials must include a degree in electrical engineering.)

The sweeping changes in broadcast technology continue—and in various ways they affect the very functions in the industry that formerly were set. The design engineer, for example, now must also be a businessman. This is especially true in the networks or at the larger stations or station-groups. Executive decisions along the upper floors of the television business are influenced and sometimes totally determined by their business impact. The determination may be their relationship to the cost of interest and the rate of depreciation. The yardstick for top-echelon decisions, then, can be the balance sheet. Many industry leaders recommend that the student of design engineering pursue also the study of business administration, usually at the graduate-school level.

Distinction should be made between design engineer and the

operating technician in the TV-broadcasting field. Unlike the design engineer, the operating technician is a maintainer. Specifically the latter's job is to maintain the newly designed equipment. This function, in fact, has changed at many of the top broadcast centers from what had been basically mechanical-oriented to electrical. There is a third category in this overview of the practical effects of the new technology. In addition to the design engineer and the operating technician, there is the operator of equipment. The equipment operator should be able to perform the responsibilities with minimal training—perhaps two years in a top-grade technical school might be sufficient for this area. It is important, however, that the operator understand the principles involved in using the new technology. As it has been expressed by executives in the field in conversations with me: "A good professional auto driver knows the engine; he's had at least fundamental training."

Julius Barnathan at ABC sees an "excellent opportunity for those who are well trained in computer technology, especially in the design and maintenance areas." Today's technology demands operators, also, who are adept in creative aspects of, say, photo journalism as a cameraman and of tape editing, as in framing and zooming.

Opportunities in Engineering

Whether it be one path or another you eventually choose, you may be asking yourself in effect, "Okay, so the field is there and I'm interested. How do I get started?" There's no easy answer to that question. Nobody can assure you of a job or of an opportunity. It's difficult and maybe impossible to point out the way, or to face you in the right direction. There are some helpful hints. They may not pay off; but your knowledge of them will make you more aware of opportunities and bring you closer to the industry.

Earlier I mentioned "the early bird" seeking opportunities as a summer relief worker. It might be useful in seeking out a summer job to circle the calendar, because it's January and not June that is the month in which to apply as vacation relief. Broadcasting goes on 52 weeks a year with no time out for seasonal changes, bargain sales, inventory or stock-taking, weekends, holidays, and the like.

The New Television Technology   41

Unlike most businesses, broadcasting, particularly in the engineering or operating services, is as busy—and as fully staffed—in the summer as in winter. So if you have some experience, say at a campus station or at a technical school or other facility that has TV engineering courses, don't wait till after the first robin of spring but apply several months earlier. Employment in a relief role can lead eventually to a more permanent arrangement.

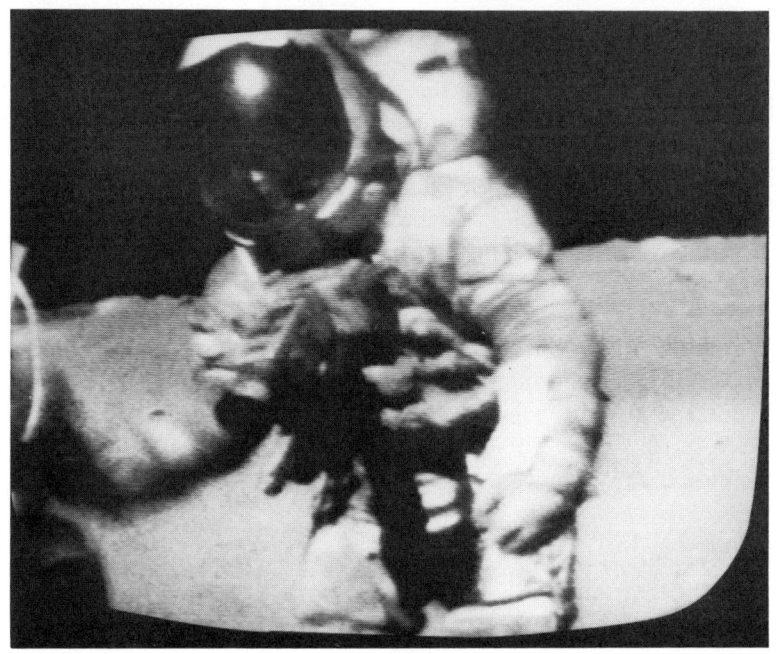

During the Apollo 16 space mission, astronauts were photographed and shown on national television even as they recorded their observations. With space exploration has come a new technology that has all but dominated the technical area of television.

Other Tips

The broadcast industry is always on the prowl for persons who are highly skilled technicians and versed in the new technology. Like what? Well, VTR (videotape recording) maintenance; or the editing, understanding of, or experience in digital theory and application. There's an interest, too, in skills, including experience in RF (radio frequency; that is, wireless radio) and, naturally, in one who has in addition to other qualifications a first-class (FCC) engineering license.

Now let's recap. If broadcast engineering is to be your cup of tea, study electrical engineering, particularly as it relates to the broadcast business. You should specialize, but it would be wise at the same time to learn all facets possible of the television business, especially those performed at the smaller, locally oriented stations. And whatever you do about this field, don't back away from opportunities to learn the new technology, nor from the computer-oriented path of study.

The future may lie with satellite technology. This refers to the fact that space technology uses digital processes (large-scale, integrated circuits) wherein information is digitalized and analoged with direct relationship to new technologies used in television broadcasting. Utilized in this science is memory storage. A single tiny chip can store 16,000 bits of information. There are no limits to the challenge and opportunity. For some, the new technology may be confounding. For others, it may very well be the New Math in their field. In any event, it is a provocative possibility for those who would explore broadcast engineering as a career.

CHAPTER VII

Sales

The broadcaster sells time. This is the way the broadcaster obtains revenue. The sale of time on the station provides the money to pay bills and salaries and permits the station's owners to reinvest in the business itself.

Sales by the broadcaster are to advertisers. Usually advertisers manufacture a product or deal in services, all for sale to the public. Although some dealings of stations with advertisers may be direct, the advertiser is almost always represented in these transactions by an advertising agency. The agency prepares commercials for TV and radio stations to put on the air, buys the exact time period of the day or the evening on the station for showing these commercials, and in return receives compensation (commission), which amounts to a specific fee or a percentage (usually 15 percent) of the fee paid by the advertiser.

Magazines and newspapers charge the reader through the cost of buying the publications at a newsstand or by subscription. They make some of their income through circulation, but the bulk of it comes from advertising. Radio and television, however, rely on advertising as their single source of revenue. Magazines and newspapers normally limit their advertising to the number of pages they can afford to print: the more advertising, the more editorial content. In broadcasting, unsold time is lost forever. There is no way to "add" a one-minute commercial to make up for a minute period that did not have an advertiser.

Advertiser revenue provides lifeblood to a station (or network), and the job of the sales department is to assure that flow. If the salespeople on the staff are doing their job and are bright, knowledgeable, energetic, and enthusiastic, the results can be outstanding. Good managers and administrators in the sales area can insure a smooth, even flow of revenue. The revenues gained make possible the entertainment and information programs that television provides.

Stations receive three types of revenue: network payments if the station is affiliated with a network; sales of time to national and regional advertisers on a "spot" basis (non-network), and sales of time locally to advertisers.

Generally, revenue comes from the networks or from advertising that the station sells through out-of-town sales representatives, both from national advertising sources; or revenue from sales to local merchants in the station's coverage area and handled by its sales force. The sales staff, in selling time periods, often has to attract advertisers on the basis of available programming and audiences. The sale of television time requires expertise, personal appearance or bearing, education, and a freedom in individual action and scope.

Television Salespeople

Educational requirements in this area are flexible. A person does not need a degree from an accredited university to sell; but he or she may need that diploma to get the job. Two persons competing for the same job will be judged on what edge is there—assuming that they are about equally qualified. That edge is often better academic credentials. Colleges offer courses in marketing, advertising, and broadcasting. Individually, and most certainly collectively, the studies can better equip the salesperson to understand and often help solve the needs and problems of clients.

Many people in the broadcasting field believe it takes a good deal of creative ability and imagination to sell television time (programs). Salaries, moreover, are quite good. In the 1960's salaries were in the

$30,000 range at many of the larger stations. The dollar figures involved in the buying and selling of TV time have accelerated. By the late 1970's, one minute of commercial time on a top show in prime time on a television network cost an advertiser more than $125,000 on the average. Television for many buyers is considered a "must" for their clients, but high priced. And, more often than not, the people who sell the commercial time to the buyers are themselves very well paid.

The Sales Manager

The general sales policy of a station is set by the sales manager, who also supervises and directs the activities of the station's sales force. Often the top person on the totem pole in sales is titled general sales manager or sales manager (in some smaller markets, it's commercial manager). At most TV stations, this person in effect serves as the second in command, ranking next to the general manager.

At the larger stations, the manager of sales has people who assist in putting together plans and "packages" (various elements of an offer which when put together form a cohesive unit or package for the consideration of an advertiser and his advertising agency). People who are so engaged in assembling the information are in sales development, new business, or research. In addition to these people—and the salespeople and managers—are the so-called middle management in sales at the larger stations, where there are also longer tables of organization. These salespeople may be called national sales managers or local sales managers, the national and local interpreting the areas of responsibility with precise labeling (an unusual situation in the broadcast business, where all sorts of titles and jobs exist and often represent different levels of activity at different companies).

The Salesperson: Personality Profile

What do we mean when we speak of "imagination" in the sale of television time? Answers will vary, but a most common response in the industry amounts to the following: an understanding of the advertiser's business requires sales imagination, and, at the same time, a willingness or a competitive drive to develop advertising approaches that appeal to the client and provide the salesperson with independence and flexibility. The sales executive becomes needed.

Some call the "something" a pleasant personality. Some tag it glibness; others, fluency. It may not be any one of these exactly, but ingenuity. Most salespeople, especially those who are reasonably successful, have that extra, hard-to-define, and nearly nebulous quality or ability. When combined with an inner drive to succeed, this quality can help a salesperson to gain giant steps in the field. But to achieve it also requires a genuine taste for hard work. So much so that those who lack a work ethic may be in deep trouble regardless of personality. Personal dynamics may be tempered by personality problems: laziness, inability to communicate with one's peers or bosses, or an insatiable desire to do other than time-selling.

The television business over the years has increased in complexity. No longer is it sufficient in selling time to be conversant with program schedules, rate cards, and audience ratings. Rating points, cumes (cumulative ratings), market profiles and research, cost-per-thousands, and packages are now familiar terms in the salesperson's lexicon.

Whole new business opportunities are developed with television. In the 1950's, for example, department stores and other such retailers advertised almost exclusively in newspapers. By the 1960's, this was changing, as more and more of these retailers and manufacturers of furnishings, carpeting, and other suppliers sampled the television advertising medium. The 1970's brought a burst of retail activity in television. Retail trade rocketed in television almost at the same time that the bright revenues of cigarette advertising burned out (cigarette advertising was prohibited on television by government regulation). Thus the withdrawal of one set of major advertisers released large chunks of broadcast time for sale. Newly developed business had plenty of opportunities to choose from—an ideal business climate for the buyer. Since that changeover, however, buyers have found it increasingly difficult to acquire prime commercial time periods, thus encouraging the sellers (the television networks and many TV stations) to increase prices.

In addition to working closely with the general manager, the head executive in sales closely collaborates with the program manager/director in developing and/or buying salable programs—that is, salable to advertisers. In view of still other responsibilities, administrative in nature (spearheading of sales staff efforts, hiring and training of personnel, and supervision of sales activities), it is not surprising that many general managers are drawn from the sales ranks.

Competent sales executives are usually in demand in TV (at networks, stations, sales representation firms, and film syndicators). Good salespeople who also have administrative acumen frequently advance to the position of sales manager.

Not everyone, however, is suited for selling. Many people enjoy more detailed, structured work. The sales departments of broadcast companies need skills for jobs closely related to the sales function. One of these is the traffic manager, a challenging role. The word traffic creates all sorts of images of activity moving in and out of the station, of mounds of papers, lines of delivery people, and the like. It is a responsible job, and the traffic manager has the backup of two or three clerks and a sales service clerk.

In broadcasting, the salesperson actually fills the function of sales consultant, working closely with other station executives and clients. This is what makes time-selling interesting but also demanding of one's creative talents.

It may be possible for you to obtain some experience in selling at a campus broadcast station or on a school paper. This may bring you into contact with people now actively in the field and eventually into a sales job. Or it may be necessary for you to settle for less elsewhere until you accumulate the experience to sell television time.

CHAPTER VIII

Promotion

No career evaluation of television should exclude at least reflection over the promotion skill in the business. Promotion functions at stations and networks today are assuming greater importance. For one thing, they encompass several distinct activities in network and station operations. For many young people, too, promotion is an area that offers an excellent working introduction into the way broadcast entities operate. Many people in TV find this path productive in vying for management positions. Many more have enjoyed the work for what it is and have stayed with it during their entire careers.

The promotion writer, the advertising person, and the newspaper writer/editor have one thing in common: they are all in the communications field. As one veteran promotion executive observed in making the point: "We grab them up front. We all try to catch the reader's attention quickly, then we'll unload the information for fast reactions and response."

As a person in the field with expertise, George T. Rodman is now a consultant for stations, station groups, and networks. He advised young people not to be overcautious. He urged them to step out immediately to seek a first job in the television business at a relatively small-market station, away from the larger cities.

Until then, he said, you can prepare yourself by taking as many writing courses as possible in school. Writing talent is essential. You should be at ease with sentence structure, syntax, and vocabulary. As a candidate for the field, you will probably be an English major, or at least a student of liberal arts. If you've earned a degree in undergraduate work, you can be expected to be planning to continue in graduate school.

People fitted for the promotion field would generally have all-around abilities but specifically would have: (1) the ability to put words together, whether inborn or acquired; (2) a conviction that people in television have to communicate in order later to be understood and that understanding cannot come about until communication is first established; and (3) a talent for persuasion and a background that includes experience with film, with the visual arts.

Check these three factors, and if you are deficient in one but still see your future in a form of promotion work at a television operation, do something about it. Become proficient where you lack skill and knowledge. Your desire must be strong enough to overcome insufficient grounding. The field is too crowded, the competition too stiff for mediocrity. Talent and preparation go hand in hand. No field is too crowded nor the competition stiff enough to beat that combination!

What is promotion? The term in the television business usually applies to a function that serves as an umbrella for several operational units. These are promotion, advertising, publicity, and public relations. Closely allied to this field (and to the news department) is community relations or community affairs. In the networks, the area is extensive, covering also corporate research, corporate information, stockholder relations, speechwriting, merchandising, exploitation (a term carried over from the motion-picture industry to denote publicity and promotion for a specific film or star), photography (studio, photographers), and other related activities.

At most stations, the job functions of promotion, publicity, public relations, and advertising are combined within the classification of promotion. At the networks or at stations in the biggest cities they are seldom, if ever, combined. At most network offices, for example, the public relations department is usually separated from the sister functions of advertising or publicity. In practice at the networks, one of these departments is often completely unaware of what the other is doing—yet the objective may be the same!

George Rodman was emphatic in his belief that young people cannot expect to progress in the general promotion field unless they have a command of the English language. Chronic misspellers need not apply.

Value of News Background

If you have a four-year background in liberal arts, so much the better and the more likely you'll have an opportunity to land a job in the field. People with newspaper backgrounds are considered favorably at the networks for placement in promotion or public relations. Generally one should be well-read, especially in current events and in the entertainment world, with emphasis on television. Rarely does a job in the field go to a person versed in such unrelated areas as accounting, law, or engineering, although all of those skills are needed in other departments at the average TV station. Sometimes the entry to promotion writing is from an advertising agency, and the reverse holds true in that a promotion expert at a station or a network goes on to become an adman at an agency.

But it is the lure of broadcasting, its excitement as an entertainment and information medium, that sets apart promotion writing and promotion work in television from promotion as performed in a nonbroadcast industry. The work may be similar, the background may need the same elements and demand the same skills, but the working climate may be something else.

Rodman explained that a newspaper background—and school paper or campus sheet would qualify in this context—can be invaluable for the aspirant applying for advertising or promotion work, "not only for the training gained in recognizing what is news but also for [news] story writing and for the ability to distribute information once it is sought out, gathered and put together."

News training can help in other ways. Rodman said that in his opinion what one does to "synthesize one's thinking for news" is the process that one similarly employs when writing promotion, an acquired skill that the promotion writer will appreciate when getting on with his or her profession. Quite often this "instant" craftsmanship is available only through experience in writing and editing for newspapers, magazines, or other news media. Again, the same process of sharpening the young person in the typical newsroom

WERNER J. KUHN PHOTO

A key role in promotion is that of advertising and promoting the fall season programming. Shown discussing layouts for the fall campaign of WOR-TV in New York are (left to right) George Snowden, program director; Bob Williamson, general manager; and Andy Duca, director of creative services. The importance to WOR of such promotion lies in the fact that it is an independent station, with no affiliation with the networks.

could be duplicated at a station or network where the writer of advertising must "grasp the salient thought" in his copy.

Promotion writers in television must be attuned to the news field, if only to become proficient in working with reporters and writers for whom they will be supplying news and information about the station or network. They must be able to write a news release. They should have an interest in community events as well as current events. They should understand how advertising works and, better still, how to direct it. All this and more. The more is essentially promotion writing itself.

You'll need to be prepared should promotion or a related field interest you. Study the techniques of film production and film construction. Sign on for programs that include them or that are related —they may be available in your school or elsewhere in the community. Learn how to assemble an on-air promo. Test your ability to write and to influence others.

If you go through college with a camera, you'll be pointed in the right direction. Launch your career preparation by enrolling in the communications department at school. In selecting a college, look for one preferably that has a campus broadcast station—assuming, of course, that all other facilities at the school are on a par with similar colleges you might consider.

To recap, the successful advertising-publicity staffer is one who has writing ability, can type with ease, reads books about advertising or marketing if there are no courses in these subjects at school, and avidly reads books about television, especially about the business of television and preferably by authors who are in the TV or advertising marketplace.

How to Get Started

You are advised to look for a start in promotion at a TV station located outside of the general metropolitan center. The best time of the year to look, the professionals say, is in the spring, well in advance of the flow of late May–early June college graduates. Rodman said he would advise young people who want to break into his field to be "prepared and to be willing to work for as little money as possible in order that they may acquire experience as an on-air promo writer or as a publicity person. It's helpful for the person involved to offer to work for little money, to work for background."

That's true about many fields. The way ahead often is based on an acceptance of menial work at the beginning. It can be true in TV work such as promotion or news writing at a station, because at many stations the new staffer has to start at the bottom. It may even prove to be counterproductive for the young woman or man in search of employment at a TV station to adopt an attitude of a know-it-all. It may be possible for broadcaster employers to allow for a lack of natural talent or a paucity of certain skills, but it is difficult for them to tolerate arrogance or cockiness.

At the networks, promotion work is available, but often the openings are quickly filled by people already in the field who are working at local stations. This happens, for example, in the larger cities in which networks own stations. Sometimes, too, the movement is reversed. A crackerjack promotion staff person at a network in New York may be hired as a director of a promotion department at the owned station in Philadelphia or Chicago or Los Angeles.

Networks, then, often hire station employees. Since networks own TV stations in the biggest cities, a promotion job is often filled by a person who first joined a TV station in a smaller community, then went on to one of the television outlets owned by a TV network, and finally moved from the network-owned station to the network headquarters in New York or Los Angeles.

CHAPTER IX

Unions

If I had to set up a few principles to guide the young American along his course of television-industry endeavors, the first career advice I'd give would be: learn to get along with other people, particularly the people you work with.

The second principle would be anybody's guess; but a lot of my friends in the industry would offer this line:

Learn to get along with unions. Know what makes them tick and how they can help (or hurt) you. Because—and this returns full circle to my first career advice—unions are people, and getting along with unions is getting along with people you work with.

Keep still another thought in mind. There is a good chance that at the time you get that long-sought-after job in television, it will be necessary for you to join a union.

It would be a waste of space in this book and of your reading time to give a detailed history of the union movement or its standing in an industry as volatile, changeable, and big-business–oriented as television.

So let's stick to generalities. We will evaluate as we proceed on the subject of unions. But let it not be said that I have a bias for or against unions. They have their place in television—and that presence is big and as important as the industry itself.

Unions

A union exists basically to *protect* or argue the interests of its members: the working force in a particular industry or craft. All of us have direct association with someone who is or was a member of a union. If not dad, then mom; if not them, then older brother or sister, or an uncle or an aunt. As a student your teachers may be union members; the postman on the block; the policeman at the corner. It has become less a question of whether or not a union

Local programming in the major cities—such as live coverage of New York Yankees baseball games—are handled by station personnel who are unionized. WPIX employs cameramen and other technicians who are members of one or more of the giant unions in the television industry.

than what type of union: how powerful? how influential? how weak? how ineffectual?

There is similarly no firm pattern of unionization in the TV industry. It is not like the automobile industry, where the pattern is set; management and union are clearly defined. In television it always *depends*—it depends on geography, traditions, and levels (some broadcast unions, like the American Federation of Television and Radio Announcers, are craft-organized, and some, as in the

Screen Actors Guild (SAG), have the word "guild" in the title).

Because unions are designed to protect members' interests, by and large they help to keep employees' pay scales at a reasonable level (reasonable in comparison with other industries in the nation, and in this context TV employees do exceptionally well); they fight zealously to guard job security, and in return they expect discipline and loyalty in the members' ranks.

Although there is no firm pattern of unionization in TV, there are the following general characteristics:

Television and radio stations in the large metropolitan centers tend to be more heavily organized than are stations in smaller communities.

Technicians (cameramen, stagehands, lighting engineer, sound man, etc.) are more heavily or tightly organized than are announcers and writers, most of whom are members of unions, whereas salespeople and programmers are seldom required to join unions.

The size of operation is a factor. The three major commercial television networks have unions. They are organized to a sophisticated degree. Unionization at the networks includes the New York and the Los Angeles headquarters operations. It includes also the owned-station operations of the networks. Each network owns five television stations, in the major cities. All three networks, for example, have stations that they own and operate in New York City; each owns a station in four other cities. In Philadelphia, for example, WCAU-TV is owned and operated by the Columbia Broadcasting System. In Washington, D.C., WRC-TV is owned and operated by the National Broadcasting Company. One of the networks—CBS—owns a station in St. Louis. All of the networks are represented in TV-station ownership on the West Coast. CBS and NBC own stations in Chicago.

Add together the five stations each network owns—or a total of 15 for all of the commercial networks' stations—plus the networks' own extensive operations and personnel, and you will find a network nucleus, so to speak, for the labor-union activity in television. Studio and film-making facilities are massive in Hollywood. The unions have historically heavily organized the movie-making industry. (A product was Ronald Reagan, who as an actor was president of the Screen

Actors Guild in Hollywood and later went into politics. Senator George Murphy of California was another actor of the celluloid capital and a product of the union activity there.) This Hollywood arm of union activity has been extended to hold TV to the banner, now that the industry has become big and powerful. Television is no longer the small-time relation of the motion-picture business. It is in the big time, and the unions intend to be right there with it.

To all this union strength, one must add just about every important—in size or earnings—television station in the top markets. These stations are unionized, with few exceptions. What's more, many small and medium-sized stations have varying degrees of unionization.

Although unions fill a role that is important to the health and welfare of the employees and, consequently, of indirect benefit to the broadcasting management, the possibility must also be recognized that unions may inhibit the ambitions of some young people eager to launch a learning career in broadcasting.

How so? To put it bluntly and perhaps to oversimplify, no matter how much help the union may provide you, the role you'll be relegated to in the industry will be governed by the union. Your activity or creativity may have to be narrowly channeled. It may be the better choice for a young person fresh out of college but nursing a burning desire to get in on the ground floor of broadcasting to hunt for his first TV job with a station that is small and not organized—or if it does have unions, only a few or concentrated in the engineering area. Such a station, if in radio, could have only five or six employees. A small television station may have only fifteen or twenty people. Not only will this type of operation permit you to expand and experiment with your talent, but also it will encourage you to do several jobs and perform all sorts of station functions. Hence you'll learn more and faster.

Areas of Unionization

Unions, of course, are not for everyone in broadcasting. Most unionization is in the technical or operational areas. Union members are usually the cameramen, sound men, lighting technicians, and other such skilled people. On still another level and in a much different area, unions include in their memberships directors, producers, and announcers. At the networks, they may include promotion writers and publicity people, newsmen and news writers, clerks, and secretaries.

But in the technical area, unless you are willing to join a union, there is a very good chance that you will remain an outsider looking in. In accepting the union, you are automatically subject to its rules and discipline.

What could this mean to you? Generally, your pay increases, your hours of work, and the specific assignments you can fill will be largely determined by the union because of rules on seniority, contents of contracts, and the like. Outstanding individual performance need not necessarily lead to promotion or even to special recognition or financial reward. You will not be pushed up the ladder quickly —that is, you won't be so pushed unless management is consulting with the union. And, of course, you may limit yourself to an area of specialization.

But there's the other side.

Your pay will be good. You'll have something to say about it possibly getting better. At least, there will be somebody to listen to your grievances, and perhaps something will be done about them. You will have substantial job security, and the union will be there to cross the t's and dot the i's.

Since many young people have been able to obtain employment in television in a highly skilled technical area such as cameraman, sound man, or lighting engineer, but without a formal, four-year academic career and with only a high-school diploma and specialized training in a technical school, the union serves as a rallying source.

Many broadcast executives in fact are former union members. They first joined the television industry in the manner described above. They started out in the technical areas of broadcasting. In many a network or large broadcast organization, a key management person will say with pride, "I'm an ex-union member." These people came up from the ranks.

There are many unions in the broadcast industry, but only a handful are of major importance. The six larger unions in the field are:

The National Association of Broadcast Employees and Technicians (NABET) and the International Brotherhood of Electrical Workers (IBEW), both of which organize all kinds of broadcasting workers including writers at the networks and major stations, but whose members are mostly technicians.

The International Alliance of Theatrical Stage Employes (IATSE) and Moving Picture Machine Operators (usually known only as IATSE, pronounced Eye-AT-SEE), which organizes various crafts including stagehands, sound and lighting people, wardrobe attendants, makeup experts, and camerapersons.

American Federation of Television and Radio Artists, which organizes many of the announcers and entertainers (and also news personnel).

The Directors Guild of America, which organizes program directors, associate directors, and stage managers.

The Screen Actors Guild (SAG), which represents the majority of talent personnel who appear on films produced especially for television.

According to some estimates, union people at one point in recent times (the scale continues to rise) were guaranteed a minimum of $165 per week and $305 weekly after three years; some members with additional experience in the TV field were being paid $100 per week above union scale.

News announcers, union or nonunion, are paid differently. They are considered apart from the behind-the-scenes personnel but nearly in the same category, at times, as talent personnel. News announcers, for example, may be paid $1,500 to $2,000 a week, though the range is far less than that approximating the $20,000 to $30,000 per-year level.

CHAPTER X

Women in Television

Why a chapter on "women in television"? Discriminatory? Perhaps. But the question about a woman's chances in television is asked often, and as a consequence we cannot ignore the subject.

Writing about careers in television for what he blithely called the "distaff side," a broadcast executive only a decade ago counseled his readers that the same positions to which a woman might reasonably aspire in almost any business were open to her in broadcasting. He specified this to mean that "secretarial and clerical positions are basically jobs for women."

As indeed they were—and in broadcasting. But times have changed.

Few women today will accept such statements. Yet we cannot be too hard on the author. After all, he represented beliefs that were prevalent over ten years ago, and in the context of that period he was on the mark. Employment of women was limited in most areas of broadcasting.

Women, moreover, still occupy most secretarial and clerical jobs in industry today. These are very much needed skills and talents. It is only realistic that the door of opportunity opens the widest for women when the search is for secretaries, in broadcasting or in other fields.

On the other hand, the woman in broadcasting is now accorded a new importance. This is so even in sales and in technical areas

NBC PHOTO

Maureen Mulvaney, a 24-year-old part-time employee of the National Broadcasting Company, worked with her father, Donald Mulvaney, in operating a minicam camera unit at the Democratic National Convention in New York in 1976. They alternated between handling the camera and carrying the pack.

where breakthroughs would have been inconceivable only a few years ago. Broadcasting, perhaps more than any other endeavor, is open to the employment and the advancement of women.

On-Camera Assignments

In the past few years, more and more women have been represented in television news departments at stations the country over.

Twenty years ago, they were a rarity in the highly rarefied air of the newsroom or before the camera and microphone in the studio. Even more dramatic than the general rise of numbers of women in news departments is the rapidity with which women have moved into on-camera news assignments, nationally at the networks and locally at the stations. In just about every major city, women are highly visible on television news shows. This seems to be part of the future, whether the station is in a small market or in the large news centers of Washington or New York, where women have become on-air reporters, anchorpersons, commentators, editorialists, and at times sportscasters.

Though perhaps the exception and deliberately glamorized, Barbara Walters in 1976 became one of the more influential, better-paid, and better-versed persons in electronic journalism. Her move from the *Today* show on the National Broadcasting Company television network to the American Broadcasting Company's early-evening network team put her in the Big Time. She captured the limelight almost overnight.

Barbara Walters had become a household name. She was now the subject of conversation at dinner and over morning coffee for millions. Far from incidental was the allocation of $1 million a year that the network budgeted for the switch.

The superstar status of a Walters could be chalked up as a talent fluke by the cynics and by her detractors. Others, however, saw in her unprecedented rise in broadcast journalism a landmark achievement in an otherwise male-dominated communications business. (Both interpretations fell short of reality.) The meteoric rise of a Barbara Walters or of a Jane Pauley (who ironically landed a top post on the *Today* show) could only provide dramatic inspiration for aspiring young women who wanted to enter the broadcast field. What then did this say for the industry and for those young women who eyed television with nervous anticipation?

For one thing, it should have reminded us that a young woman may still be hired in TV as a secretary and be assured good pay and a pleasant and rewarding career, if that's her hope. But it also told us that TV has dropped its off-limits signs so far as women are concerned. It told us that the industry no longer would abide by *for men only* attitudes or beliefs. One's sex, then, could no longer provide the rationale by which the male television businessman could perpetuate the myth that only he could perform with professionalism.

I recall that only a few years ago the three commercial television networks purposely set out to find women who could fill sales positions then occupied on staff level only by men. NBC has figures

NBC PHOTO

Jane Pauley of NBC shot to stardom in news after having been "discovered" while broadcasting the news as a co-anchorperson on a Chicago television station. She is one of a growing group of highly paid, talented, and successful women who appear in front of the television camera.

showing that the number of women employed in sales shot upward from the point of none in 1970 to an eyebrow-raising 17.2 percent by 1976. Interestingly, although the sales ranks were bereft of women in positions of responsibility at NBC, by 1976 the percentage of women employed there in sales was higher than in any other area at the network with the exception of what NBC classified as "professionals and administrators."

Increase in Hiring

Computations by NBC showed that women represented 16.2 percent of the network's staff positions in four groupings: officials and managers, professionals and administrators, technicians, and sales personnel. Six years earlier, the percentage of women was but 6.4.

In addition to the increase in the hiring of saleswomen, the percentage of officials and managers who were women doubled at NBC from 6.3 percent to 13.9 percent in the period. The number of women in the category of professionals and administrators nearly doubled, too, increasing from 14.4 percent to 27.9 percent. The number of women technicians went from 19 (or 1.2 percent) in 1970 to over 100 in 1976, an astounding exposition of change in the role of women in broadcasting!

What these trends of the 1970's indicate, aside from dramatizing the obvious change in the woman's status in the broadcasting field, is that assumptions must be obsolete. For example, in the estimate given in the early 1970's that women then represented 21.1 percent of all on-air "broadcaster" jobs must have radically changed upward by the mid-1970's and even more so from the time these sentences are being written to the moment you'll be reading them.

A quick check in the summer of 1976 showed several women holding top executive positions in the networks, a half dozen vice-presidents or their equivalent in rank at the American Broadcasting Company alone. Statistics on women in broadcasting are no sooner compiled than they must be revised and updated with higher numbers.

Most people in the broadcast industry believe there is a better climate for women in broadcasting than in other businesses. Be that as it may, there is constant effort within the industry to upgrade women's role. The Columbia Broadcasting System for years held seminars, counseling services, and training activities in career development for women.

In the 1976 Presidential debates on television, three women network on-air broadcasters—Barbara Walters of ABC and Pauline Fredericks and Marilyn Berger of the National Broadcasting Company—represented the broadcast industry. Six television news personalities in all were represented in the debates (including the debate between Vice-Presidential candidates Mondale and Dole). In contrast, there were ten newspeople from the printed press, of whom only one was a female.

Getting Started

A foremost personnel expert in the TV field is Sherlee Barish. In an interview at her New York office, she acknowledged that although television may be loaded with opportunities for women, "they [the industry] don't want anybody without experience. Once the applicant comes up with job experience, the opportunities materialize."

The opportunities in news work in the top 75 markets, for example, exist for women in on-air work, reporting, news production, and anchoring. Yet, she said, because television was a "glamour industry and over-manpowered," there was no problem in recruitment. "The best bet," she advised the woman who is ambitious in television work, "is to look for a job in the smaller television market where you will be in demand. You'll rate [that is, get an even—or better-than-even—break with a male], both on the air and in behind-the-scenes work."

Specifically, Miss Barish said, "there is a wider open door for women. But they must get background. Without it, too, obvious opportunities in television dry up."

When Miss Barish referred to a "smaller market" below the 75 top markets, she had in mind a market the size of a Medford, Oregon; a Madison, Wisconsin; an Altoona, Pennsylvania; or a Savannah, Georgia. In these areas, according to Sherlee Barish, opportunities in television management have been also more readily available to women.

Miss Barish observed that during that very week in the fall of 1976 when we had our discussion, the position of news operations manager opened up in Washington, D.C., for the right woman. She said, however, that the job was hard to fill, principally because requirements (weighted toward experience) had to be met, a concern common to the largest TV-news markets.

Whereas women in many fields consider the employment practices in their individual industries to be set practically in the Neanderthal

Age, women in television are usually pleasantly surprised by what they find in their field. As Miss Barish noted, the pay in the news area is about equal for men and women. Conditions in the broadcast industry for the most part are naturally encouraging as well as tolerable for women. This is so not only in broadcast news but in just about any aspect of the television business.

Although women are achieving near equality with men for employment opportunities in broadcasting, there are still roadblocks to get through before they can expect to obtain a footing in the industry. For one thing, women, not unlike their opposite numbers, must contend with broadcasters who traditionally tend to promote from within. Because of this, the odds are automatically increased against the so-called outsider, regardless of sex, who seeks a first opportunity in television.

Another block is the unofficial stand at the networks against people "using" an educational degree—particularly in "broadcasting"—when seeking employment at station or network offices and studios. Networks as a rule don't like broadcasting courses or degrees. Experts in personnel practices who are candid will advise job hunters in news, for example, that they are best prepared if they have a degree in journalism, or that they graduate with a liberal arts background, if not a degree, or that they be a history major—almost anything but a graduate degree of broadcast studies.

To take the long view: the ball, so to speak, has been lobbed into the woman's court in the field of broadcasting. In the mid-1970's the opportunities for jobs in the broadcast industry are no better and no worse than they may be in other businesses. But this you can count on: At no time in the history of television has the opportunity been as good as it is now for women to make their careers.

Unfortunately, it is not possible to equate opportunity for jobs with opportunity for sex equality. Though you may be treated as a woman as fairly as the person of the opposite sex and given equal consideration, there may not be a job for either you or the man who has been competing. Or it may be that job experience—even a lot of experience—is a prerequisite that counts more than education or sex.

CHAPTER XI

The Dimensions of Television

Television *is* big business.

There are more than 700 television stations operating commercially in the United States. Three TV commercial networks provide interconnected services. Each of these networks owns, in addition to radio stations, five television stations, all in major cities. Most of the commercial TV stations, moreover, are affiliated with the networks. Roughly speaking, each network has some 200 affiliated TV stations. More than 100 commercial TV stations are independent in the sense that they are neither owned nor affiliated with any TV network.

In the mid-1970's revenues in television had topped the $4 billion mark, and, according to Federal Communications Commission figures, profits were over $780 million. According to trade estimates, the average 30-second prime-time network television announcement cost the advertiser $50,000. Since many spot commercials are paired as a 30 back-to-back with a 30, or combined as a 30 with a 20 and a 10-second commercial, the common one-minute commercial was bringing in about $100,000. One-minute announcements were increasingly costing more. Adjustments in network pricing of prime time were underway in 1976 and into 1977. A one-minute announcement in the football Super Bowl classic on television cost an adver-

68   YOUR FUTURE IN TELEVISION CAREERS

tiser $250,000 in 1977! The return in audience, however, was impressive: an estimated 75 million people viewed the televised game.

Other statistics help in putting together the dimensions of television. Over 70 million homes have television sets, and nearly half of them have more than one set. More than 50 million color sets were

Television's area of interest is as expansive and as deep as society itself. One of its more dramatic episodes occurred in the fall of 1976 with the television debates between Presidential contenders Jimmy Carter (left) and Gerald Ford. Special care was exercised to provide each candidate with equal facilities. This was the second time in history (the first was in the early 1960's between John F. Kennedy and Richard M. Nixon) that Presidential candidates debated the issues before a national television audience.

included in these figures in 1977. The average family watched television a little over six hours a day.

The largest states, by either population or area, have the most TV facilities. Including all TV stations, California led the states with 75, followed by Texas with 68. Another sun state, Florida, had 42, and New York had 40. Rhode Island had but four, Delaware one, Utah and Vermont five each, and Wyoming three.

Large stations established in metropolitan areas of the country employ as many as 250 people, according to estimates supplied by the National Association of Broadcasters. Smaller stations in smaller cities have closer to 30 people.

Sources of Programming

There are distinct sources of programming on commercial television stations. Local shows are created and produced by the programming staff of the station. They produce children's shows, religious programs under the auspices of local churches, interview or panel shows, sports coverage, public events (parades, for example), and most important, news and news documentaries. Independent program producers and syndicators provide programming, including motion pictures and such varied fare as wildlife shows or game shows, and dramatic series of all kinds including some formerly shown in network prime time and now classified as "off-network." For the network-affiliated station, a large part of programming comes from the networks themselves. By contractual agreement, affiliates agree to carry programs supplied by a network during the broadcast day. The network produces and sells time in the programs to advertisers, and shares these revenues with the stations through compensation.

By now, you must have a picture of television that is different than your earlier one. Images on the screen still come to mind. But you must also see in your mind's eye cameras, studios, sets, buildings, offices, people interacting as well as stage acting—a myriad behind-the-screen images. Perhaps you have had actual experience in broadcasting at your school, and as you read these chapters you feel at home, as though this *is* your industry.

Not included in these dimensions are the extensive offshoots of the industry: public (noncommercial) television, cable television, closed-circuit television, and pay television. These are summarized in a later chapter.

By now also you are aware that commercial television is an

interface operation of stations and giant networks. The preceding chapters and particularly this chapter were designed to tell you a little about every aspect of the industry. Most of what you have read has focused on several functions, including administration or management, programming (including news), engineering, and sales. To some extent, the spotlight was deliberately diverted from the more glamorous networks to stations. This was necessary because the stations are basic elements. Without them, networks as we know them would not exist.

For the remainder of this chapter, then, let us concentrate on networks and station groups, to gain additional information as to their makeup and function as well as job areas.

Networks

In any study of networks and available jobs, one factor stands out quite clearly: the network world demands experience before there is advancement; talent and determination before there is development.

Much of what a network is, does, or plans usually holds true for television stations. Basically they are involved in the same activities: the production and distribution of programs and the sale of time to advertisers. Networks have all the basic jobs found in stations and then some, because of the complexities of the networks' programming activities and their operations. A station houses all staff and most facilities under one roof in a single building. A network operates several facilities, owns stations, and has offices, studios, and the like in more than one city and in more than one location within a city.

Key program personnel at the networks plan the network schedule, and members of the department are assigned to specific programs, projects, or series. The team is made up of managers, producers, directors, writers, musical directors, announcers, costumers, scenic designers, and production assistants. This team is considered to be a production unit with the addition of performers, the technical and stage crews, makeup artists, and production coordinators. Many of these functions have already been discussed in preceding chapters. In any event, they cover areas wherein careers are available in the broad TV spectrum.

Many of the same production jobs are available in network news and public affairs as on network entertainment shows. The department also has news writers, researchers, commentators, and, of course, foreign and domestic correspondents.

Network sales duplicates the general function of selling time on

The news panel at the second candidate debate of the 1976 Presidential election. The panelists are (left to right) Pauline Frederick of National Public Radio moderator; Richard Valeriani of NBC News; Henry Trewhitt of the Baltimore Sun, *and Max Frankel of the New York* Times.

the station level, on a wider scale. Network salespeople call on large national companies and their advertising agencies. Network commercial operations are interwoven (or at least involved) in the planning and execution of national sales campaigns. Supporting this effort, in addition to sales personnel themselves, are the publicity and research departments.

Public relations, promotion, and advertising, in addition to research and publicity, are supportive of all key areas including top corporate offices, programming, sales, and engineering.

Engineering at the networks requires extremely well-trained and skilled people who are competent in the operation and maintenance of advanced equipment. A number of network engineers additionally work in experimental and developmental projects.

Station Groups

The station functions of staff involvement with the network and sales representative who is out of town are nominal and limited. But networks have full departments staffed to handle complex relations with affiliates, which include agreements to carry programs, clearances for shows, and technical arrangements. The networks, as mentioned previously, own and operate their own radio and TV stations as a division of network activity. In effect this activity is considered a station group. Group owners, other than networks, include such large operations as Westinghouse Broadcasting Company (Westinghouse Electric is the parent), Cox Broadcasting (initially also newspaper interests), Capital Cities Communications (also newspapers), Storer Broadcasting, and Metromedia. Of interest is the fact that most of the stations owned by the groups are also network-affiliated. The groups, which are usually set up to duplicate network and station functions of administration, programming, sales, advertising and promotion, and research and development, offer job opportunities accordingly. They also operate their own companies in related areas. Examples: Westinghouse Broadcasting has sales representation companies and a program production firm, Group W Productions; Metromedia similarly has sales rep companies of its own and Metromedia Producers Corporation, a program production company.

Normally the groups operate stations in various cities around the country. There is no special pattern. However, each of the groups has headquarters in a different city and in different regions of the country. Several happen to be in New York: Westinghouse, Capital Cities, Corinthian, RKO General, and Metromedia (also in Los Angeles). Others are in the South: Storer Broadcasting in Miami,

Cox in Atlanta, Multimedia in Greenville, South Carolina, and Wometco in Miami. In the Midwest, Taft is in Cincinnati, John E. Fetzer in Kalamazoo, Michigan, and WGN-Continental in Chicago. West Coast groups include among others Kaiser in San Francisco, King in Seattle, and McClatchy in Sacramento. Upstate New York major groups include Park Broadcasting in Ithaca and Newhouse in Syracuse. Post-Newsweek Stations has its headquarters offices in Washington, D.C.

The list of groups above is not meant to include all major station group owners in the country. It names but a select few. The intent is to show the diverse regions and to indicate that the numbers of stations and station groups, the services they offer, and the employees they have on the payroll are quite extensive.

It must be stressed that the networks of themselves are big-business enterprises. Take but one commercial TV network—the American Broadcasting Company. ABC operates a TV network and a radio network, sales representation companies in both radio and TV (ABC Radio Spot Sales and ABC Television Spot Sales), a TV-station group (ABC Owned TV Stations), a radio-station group, a programming unit (ABC Entertainment), and individual stations (five in TV alone). In addition there is ABC News and ABC Sports, both large and vital departments under the network umbrella. Top officers at ABC are paid in the hundreds of thousands of dollars a year in salary. CBS and NBC are basically similar in the subsidiary broadcast operations. Often, particularly for purposes of special news-event coverage, the networks rely on affiliated stations to provide facilities and personnel and air their reports.

CHAPTER XII

For Those Who Manage: Dollars

Each year the Broadcast Information Bureau, Inc., in New York, publishes a survey of "personal finances" in the television industry, with emphasis on salaries of top executives. The survey concentrates on station managers, program directors, and sales managers; it is intended to present a general picture of what these people are paid, with salary ranges, increases over the previous year, and the like. The most recent survey available to the author was that conducted in December 1976. The following summarizes some of those findings:

The Station Manager

This traditionally is the top job in the TV-station field. Accordingly, it is also the highest paid. The "typical" station manager, who was described as in the employ of a network affiliate in a market below the top 10 but above the 100th market, made $50,000 in 1976. The average income was $41,355. The salary represented a 10 percent increase over 1975. Income was also up 10 percent from $38,603 for 1975.

In 1976 more than 56 percent of the station managers questioned earned over $45,000 a year. This compared with 45 percent in 1975. In 1976 more than 37 percent earned over $50,000, an increase from 21 percent reported in the year before.

The Sales Manager

On the average, it was found, the station sales manager—holding the next to the top position at the typical television station—did not earn as much as the general manager.

Accordingly, the survey reported the sales manager's earnings at an average $37,311 in 1976, a *20 percent increase* from $31,000 in the previous year. There was a reason for this, however. The year 1976 happened to be burgeoning with time sales, increased rates, and an increased inflow of dollar volumes. The group of sales managers indicated that they expected a 10 percent rise in 1977!

The typical income of the sales manager was up 10 percent. Of those surveyed, 39 percent earned more than $45,000 in 1976, and nearly 17 percent earned $50,000 or more.

The Program Director

Most of the program directors surveyed earned around $20,000 a year in 1976, about 3 percent earned more than $40,000, and nearly 1 percent made over $50,000. Average earnings increased to $21,638, compared with $20,172 in 1975. The survey found 46 percent of the directors earned between $17,500 and $22,500, and nearly 7 percent earned more than $37,500.

Taking the surveyed group as a whole, it was found that only about 1,720 people in the U.S. fitted into the general/station manager, program director, or sales manager categories. The whole group earned $48,350,000 in 1976, or an average of $28,126 per person.

The startling rise in salary, particularly in the station manager group, was charted by the survey in comparing 1973 with 1976. In that relatively short span of years, the percentage of station managers earning $50,000 or more a year shot up from nearly 16 percent in 1973 to 37.3 percent in 1976, or a more than doubling of the percentage! Also of interest: in 1973 survey findings showed 15 percent of station managers earning between $20,000 and $25,000, but in 1976 the percentage had dropped to about 9.5 percent. Again on the higher scale: approximately 19 percent of station managers in

1976 earned $40,000 to $49,999, but only about 15.3 percent in 1973.

The percentage gap in higher salaries paid program directors in 1976 started around $19,000 a year. That is to say, the percentage of program directors receiving that salary or more was noticeably above 1973, but under that level the opposite was true.

CHAPTER XIII

The Noncommercial Station

The big volatile business of television is in commercial operations. So it follows that that is where the jobs are.

Advertising, we have seen, sustains commercial TV. It's the lifeblood for both station and network. Without it, the great arteries of the television networks would dry up. Television would no longer be in the competitive, driving, and creative position that it is in today. Yet there is room for other than commercialism in television.

That room is being taken up more and more by public television. Perhaps no medium has had a longer, harder row to hoe. Over the years ETV (educational television) has had to struggle, first to survive and then to grow slowly and with fits and starts, error and miscalculation.

Public television has come a long way. Today it has network facilities as in the Eastern Educational Network, the Corporation for Public Broadcasting, and the Public Broadcasting Service, each providing facilities and/or programming. Educational institutions, national foundations, and strong local groups, as well as federal government agencies, number among the funding sources in the field.

Although public television is in a steady growth pattern, nearly all the more attractive and best-paying jobs in the television industry are available in commercial television—the networks, the stations, the motion-picture studios, and related areas. Public TV then must be considered in the light of offering additional opportunity for those who would enter a television career.

Where are the opportunities in public television? The technical facilities, of course, resemble those in commercial broadcasting. Technicians man the same equipment, though perhaps it is more

CARL SAMROCK PHOTO

A familiar series to public-television audiences in 1976 was The Adams Chronicles. *The series of thirteen weekly one-hour programs traced 150 years and four generations in the lives of the Adams family. Here, John Adams (George Grizzard) is sworn in as second President of the United States. At left are George Washington (David Hooks) and Thomas Jefferson (Albert Stratton).*

limited and lacks the capability of that at the commercial networks and stations.

Public television has had a couple of strikes against it from the start. To be as productive as commercial television, it would have to outman and outsell the competitors. But that's not the way it works. Public TV is not in town to sell to the same or similar clients of commercial TV. It's there to supply an alternative television service to what the networks provide in the manner of entertainment, news, and public affairs. Public television has met the challenge by providing a news-in-depth program as contrasted to the usual news report put together by each of the commercial networks; such dramatic fare that is not normally to be seen on the giant TV chains; children's learning programs; panels; interviews; cultural series; and such sports matches as soccer and tennis.

Funding for ETV

Since ETV does not carry commercials, its stations must obtain funds from independent or governmental sources. Congress has been a benefactor, but so has the Ford Foundation. Other such strange teammates in the TV field are to be expected. Many sources have been representative of funding but with difficulty, though this problem is fast being ameliorated. Still several novel and voluntary funding sources are active in the field: by subscription or membership in the station, on-air auctions, special donations, and telethons and talkathons. Many of the educational television stations—WNET in New York, for example—employ all alternative ways for raising funds including those noted above.

In 1977 the number of public-television stations in the country was over 250. Public broadcasting, moreover, had an income of $346.8 million in 1975. Of that amount, 25 percent was estimated to have come from the federal government.

Cultural Role of ETV

Public television in this country is designed to serve the educational and cultural needs of the community. That is how the public-TV stations got started. Opportunity was provided by the government (the Federal Communications Commission) in the form of assigned educational channels. Stations were reserved, and by 1952 it had assigned noncommercial channels to 242 communities. In 1960 the number was increased to keep pace with the overall rise in the availability of channels.

Television refuses to be classified, defined, or generalized. At the same time, the medium is versatile, pliable, flexible, all-pervasive, and

Public television can be a learning experience for preschoolers who watch Sesame Street, *now a staple show provided by the noncommercial television service. The Cookie Monster, a fuzzy concoction, is a favorite with millions of children.*

at times—in view of content—almost mercurial. Neither are its forms ironbound by definition or use. For example, public TV may actually be licensed to a local educational center; many ETV stations are, in fact. At the same time, schools affiliated with that center may have closed-circuit systems for classroom use, and those sets will pick up the public-TV channel and use it during specific periods of the day. Other systems in the same educational complex may be fed programs through a wired or cable system. To confuse the issue even more, some colleges and community groups operate TV channels *not reserved* for educational purposes; thus they can realize a profit on the station's operation by carrying commercials, but they need *not* do so and may operate on a nonprofit basis.

Getting into the Field

Educational television has been in need of young people with stimulating ideas. Among other attributes cited are determination and dedication. But this enthusiasm and drive may not be enough to sweep you through the employment gates and up to the front door

PBS PHOTO

A function of public television is to provide gavel-to-gavel coverage of such events as the Watergate hearings in Washington. Shown here during a hearing of the Senate Select Committee on Campaign Activities are Senators Howard H. Baker, vice-chairman, and Sam J. Ervin, chairman, and their chief counsel, Samuel Dash.

of ETV. Openings in that field are no easier to find than in commercial television, motion pictures, radio, cable, or pay TV.

As in commercial television, you need preparation and training in public TV. Many enter the field after having a solid liberal arts education and some experience in a specific field. If that area can be television, then you are in a good position. A summer job in almost any capacity with a local station would be helpful in relating you to the industry. School courses per se may not be enough to round out your experience for TV work. Write for the school paper, join the dramatic club, the glee club, the campus station. Watch TV and evaluate the programs, consider volunteer help at the local television station. Try your hand at helping the station during fund-raising time. Some of the projects may put you in the position of learning at firsthand the problems of local ETV.

In educational television, the positions of general manager, program manager, and production manager are the critical ones because these people plan and direct the execution of the program service. A problem in public television is the lack of fully qualified people (people with the understanding and comprehensive outlook, awareness, and interest necessary in a television outlet that is interlocked with the educational process and institutions).

If you are determined to start in public television, irrespective of the employment problems in your area, you may have to settle for such jobs as sweeping the studio, filing records, typing logs, and even serving as a "go-fer" by running out for coffee and doughnuts for the program staff.

Possibly with few insiders realizing it or giving it a lot of thought, public television has taken off, after having sprouted in the 1950's and added growth in the 1960's. The expansion has boosted public TV today to a point at which it is visible, viable, and, for television, a natural job mart.

CHAPTER XIV

Widening the Field of Vision

Have we finished with career opportunities once we've discussed TV broadcasting, commercial, and public operations? There should be more than only a smidgin of opportunity in what we might call (for the purpose of this chapter) the leftovers. Or, the by-products. There is also the fact that although today most opportunities for careers exist in television broadcasting, TV is an industry of change and of dynamics. Television development cannot be predicted or predetermined. Even the TV "mood" of a country can change overnight. As a business, too, TV is fast-paced, sensitive, and downright impossible to evaluate. Guessing is nearly worthless. Even the expert prognostications in TV are often discovered to be inadequate, or falling behind the rapidly adjusting and changing society in which TV takes on the role of leader for a time and then falls back to become a follower of the pack.

We have said that commercial television is where the money is and where the career opportunities lie. Now we'll add this: it's possible, too, that the whole scene will change rapidly or abruptly, and the jobs will be opening up in the so-called nonbroadcast TV operation.

Not that cable television or pay television or closed-circuit television are closed to career-minded young people. They do have openings. But not to the extent that commercial networks or stations or public television do. They haven't expanded that much; they are not of the size, and the activity and coverage are not comparable.

In looking at the fields in television, it would be helpful to describe each one. Public TV uses the airwaves but without commercials (advertisers are permitted to "underwrite" a program; in effect they finance the show and their name is flashed on the screen). Subscription or pay television can be broadcast or sent on a cable or wired system. Closed-circuit television can range from a small, vest-pocket-sized operation up to a highly sophisticated package of multiple use.

Pay television may include a device that when activated by the viewer unscrambles the signal to deliver a clear TV picture. Cable is wired into the home, and the householder is charged for the service. Closed-circuit possibilities are extensive, since they depend on access rather than subject matter.

Closed-circuit use includes classroom teaching, factories, service businesses such as insurance agents, security systems, and even monitoring surgery. Like closed-circuit TV, commercial broadcasting can be used in different ways. In fact, the uses of television are virtually unlimited. They are limited not by the technology but by the economics of operation. You can make use of television in almost every conceivable way, but somebody has to show you how to make it pay.

Cable television is still in a period of growth and expansion. It is difficult to predict how important this segment of the industry may become in the general scheme of things in the job market. A complication is the fact that cable television usually requires no more than a few people to operate a system that covers a relatively large segment of country and of population.

Cable Television

This is generally how cable television operates: The cable system picks up a TV signal—a program transmitted by a commercial station—amplifies the signal, and then distributes the improved signal (with improved reception) to the subscriber for a prescribed fee. Simply put, cable television is a relay.

Cable has become a contender of sorts in the TV business. It is particularly effective in those areas where ordinary video reception is spotty. These include rural areas, far from the city line and away from the usual areas of reception, and perhaps on the "wrong side" of a high hill or mountain; or city areas where skyscrapers interfere with TV reception. Cable also appeals to viewers who are within a reception area but subscribe in order to receive additional channels that provide programming not usually available on conventional television.

Cable TV was "on the scene" by the early 1950's, at which time it provided service to those communities unable to receive TV signals adequately because of terrain problems or the great distance between the TV set and the point of transmission. Since those years, the concept of cable has remained quite the same, but its possible effects on community life-styles and interests are changing and hold promise of an impact not conceived of a few years ago. It should be obvious, too, that any industry that is undergoing even moderate growth should be receptive to young people seeking career opportunities. Cable, albeit in spurts, is undergoing continued expansion.

A key to the future viability of cable is resolution within the industry of its programming problems. Cable operators would like to be able to program fresh, newly developed material, including first-run feature films and sports events as they happen. Commercial broadcasters resist this. They feel that the removal of all restrictions on what can or cannot be picked up for use on cable will give cable firms carte blanche on programming and also permit them to charge the viewer according to what the traffic will bear. Thus, they argue, the viewer will be taxed for the right to see what ordinarily would have been provided free of charge by the commercial station that asked only to be permitted to play commercials adjacent to or within the program.

Whatever the resolution of such problems in cable, pay television, or public television, commercial television in the foreseeable future should continue to provide the most jobs at the highest level of salaries in the communications industry. It is not quite clear what effect these other modes of television communication, now small and relatively unimportant or ancillary to commercial TV, will have on society and on the communications industry itself once the stops are off and expansion is accelerated.

There are a number of cable systems in the United States. They developed over the years into a busy set of operators, with many new entrepreneurs entering the field. In the late 1970's there were about 3,500 operating systems serving 7,800 communities. Additionally, there were 2,650 systems approved but not built. Although cable systems did not appear to be in a position to establish a dominating stance in the TV business, the industry continued its expansion.

Cable operating systems were already serving about 10 million subscribers in the late 1970's. These subscribers translated to more than 30 million people, or 15 percent of the nation's television households. Most systems now offer 12 channels, but this number will climb presently to 20 channels. Monthly fees range as high as $9, and installation ranges from no charge up to $100.

According to government figures issued in mid-1977, cable television's revenues were over $894 million in 1976, pretax net income was slightly over $27 million, and the assets of cable operations were $2 billion.

CHAPTER XV

Attitudes and Attributes

It takes more than talent to get into television—and, once in the field, to make a go of it. The TV industry also demands specific attributes and general attitudes. It is what the broadcaster, who will hire you, thinks is important that really counts.

The broadcasters want good people, of course. They want honesty and loyalty, a desire to work in a business often described as creative, fast-moving, and "much in the public eye." But irrespective of the position you're vying for, all employees in the broadcast field ideally should have several of the following "special traits"—as cited by the National Association of Broadcasters in its "Careers" pamphlet for aspirants:

"Enthusiasm . . . a valuable asset in a changing, dynamic business."

Indeed, this must be one of the traits. Without an innate "up-and-at-'em" attitude, you will be hard put to land a job in the business, much less advance in it should you manage to be hired. Enthusiasm can be infectious. Since so much of the broadcast business is based on talent plus teamwork, enthusiasm and a willingness to try something new, while tailoring your individualism to the general objectives, will go a long way toward making a career in television a rewarding experience.

"Sense of public relations . . . no industry operates in a goldfish bowl more than broadcasting. Every employee must be conscious of the effect of his activities, both during and outside working hours."

This doesn't mean you must be a prude. But you cannot afford to be in the broadcasting business and constantly make a boorish fool of yourself among your neighbors or among your professional peers. In the positive sense, you must be prepared to become a solid citizen of your community and of your country. A news editor is no less expected to be a person of integrity, intellectual honesty, and community leadership. The broadcaster thus must assume a role and live it to the hilt.

"Creativity . . . the essence of programming is creative effort, and everybody in broadcasting can make a contribution to this important phase of the business. But creative people must learn to create cooperatively and must accept the fact that their ideas will be changed and melded with the ideas of others."

So we're back to where we started in this advisory. What this means is teamwork. Harry Reasoner said it about news; the industry's career pamphlet goes further and makes it all-inclusive—in the whole programming area, not just news. Many good ideas have been hatched in the sales department but mothered and nursed and nurtured in programming by all levels of creative people, young and old, the inexperienced and the "hacks." Broadcasting is a hungry enterprise. It must be fed constantly, and the pinch on creative minds and talents can be horrendous.

"Balanced temperament . . . broadcasting is show business with a stopwatch. It is geared to quick decisions and quick action. Employees need the ability to perform work quickly under pressure. There are some jobs for the contemplative person, but not many.

Broadcasting needs people who do not get upset easily and who can work well with others."

Any person who has turned out a piece of prose he thinks is anywhere from presentable to a great work of art and then watches an editor wack it into shape—but no longer resembling the original—knows the feeling, and the meaning of "working well with others." Similarly the filmmaker who has ground out cans of film footage and then watches the cutter-editor snip away until most of the film covers much of the cutting-room floor knows the feeling.

Similar sensitivity ought to be left at home before you enter the station or the studio. Artists who are creative are often sensitive. But that's not the type of sensitivity we are talking about. It is rather the person who lacks talent or whose talent has not yet ripened but who is sensitive, resentful, and slow to cooperate and learn his mistakes or his failings.

"Reliability . . . the broadcaster sells time. Once gone, it is gone forever. It cannot be warehoused and sold next year. Thus, meeting deadlines and getting a show or announcement on the air properly is vitally important. The entire effort of a station depends on the people on duty at the moment something happens."

Another word for this trait is *responsibility*. This is something that I would emphasize in counseling anybody who wants to work and prosper in broadcasting. A young man or woman who lacks responsibility is just another unnecessary applicant. This is one attribute also that I would recommend you *illustrate* the best way you possibly can in your résumé. If you can show you've been the person who has received the confidence of your peers and your superiors in whatever the endeavor, you'll have a good chance of attracting the attention of a future employer.

Time then is fleeting. For the broadcaster, there is never enough of it. Writers, editors, programmers, salesmen, directors, producers, clerks, secretaries, technicians, managers are all in the same boat when they work for a broadcast organization. They must work for, with, and against time. Time is their nemesis; time is their friend and savior. One thing you can do even before you start your career in broadcasting is to learn how to deal with and conquer time. Find out what it is to work against a radio or television deadline.

"Initiative . . . this trait is desirable in controlled amounts. An inventive and improvising individual can be a great help in a broadcasting station, but if he plunges ahead impervious to other departments and other people, he is often ineffectual."

What's more, he can be fired. That is the way the broadcasting shoe fits. If you wear it with patience, the shoe will last you for a long time. Try to tap dance with it in a studio when you should be soft-shoeing it in the library, and the shoe is bound to pinch as you are hastened out of the building. If all this sounds too murky to follow, let's get down to the hard facts. People in broadcasting, no less than in other industries, like to see people with some nerve and verve. But they don't want to be overpowered and "advised and counseled" by their junior employees, especially when the advice and counsel are not being asked for.

"Business sensitivity . . . broadcasting is supported by the sale of commercial time to advertisers, and the job of advertising is to sell goods and services, a fact that the alert employee keeps constantly in mind."

To anyone who has never worked at a radio or television station, the last statement may seem slightly crass, commercial, or at the very least, insensitive. That belief could not be more erroneous. It doesn't take an employee long, once on the job, to know what the objective of management is in the broadcast industry. People who work for management may have other than a profit motive uppermost in mind. But management must sell time to advertisers or else get out of broadcasting and into something else.

My first job in the business was as a reporter-writer in the news department of a small radio station in the metropolitan center of Washington, D.C. It was not long before I realized the importance of the seemingly natural acceptance afforded to me in those formative years. Management almost instantly regarded me as a responsible, reliable, and sensitive person who would do well for my immediate boss, the news editor; for my superiors (almost everyone at the station was included in this category); and eventually for my news sources.

One had to be invariably cognizant not only of what was doing out there in the world (since news events were to become the name of the game to me), but also of what was happening inside the "real world," the radio station I worked for.

I was given a full dose of responsibility. I was at work at 6 A.M. The only other person at the station was the radio personality who worked from midnight to eight in the morning—the station operated around the clock, 24 hours a day, seven days a week, week in and week out, year after year.

In effect, then, I "opened" the station, because the programming consisted of music and news and I was "Mr. News" at six o'clock. Without my cleaning up the news wires and immediately knocking out five minutes worth of opening local news, the station was only half in order and only half way toward operating. The choice of news was mine. The writing was up to me. I was expected to make instant judgments and to write that fast, or faster if need be.

If I was to be aware of anything at all, it was the importance of that commercial log. That was sacrosanct. The man at the turntable (the engineer who was also producer and director in those hours and in those years) had priority on when and how he was to play the commercials. And it was the selling proposition that proved my undoing and placed me at least for a while on the roll of the unemployed. I lost my job when the station had to cut the payroll and economize when it lost the Chesterfield cigarette sponsorship of the baseball games. That was a big contract. It sold a lot of time. When Chesterfield pulled out, the station for a period was left with lots of unsold time, and I became expendable.

So in one sentence: One must keep constantly in mind that broadcasting is supported by the sale of commercial time to advertisers, yet one must not get upset easily, not even over the fact that one is reliable and responsible, yet "expendable."

If that statement sounds rather strong, it is meant to be. But with tongue in cheek. One must allow the remembering to linger a while longer if there is a message to be learned and passed along.

Stop. You've read this far. Throw up the book and let it come down again and read on—and into a new set of "traits" as expounded at one time by another person who had collected them and written them down for posterity, or better still, for your benefit.

In this new listing one finds the word *perceptive*. This person would expect the young applicant to have the ability to quietly *explore human relations* problems. Conflicts between people one works with should be instantly recognizable. As a broadcast worker you would be a *doer* who wants to get things done, happy when challenged and impatient with the status quo. You should be willing to *sacrifice,* to want to advance and like hard work, but to dislike the routine. You should enjoy the "challenge of human relations," be an organizer, like to take charge of others, put personal performance above security. Most of all, you feel that responsibility in business is a challenge, you are quite well grounded in most areas of art, literature, and economics in addition to history, political science, and other liberal arts subjects, you are *enthusiastic* and *accept* change.

CHAPTER XVI

Preparing for Careers

At this point I must assume you have at least an appetite for work in the television industry. Perhaps you haven't come to a decision as to a specific area of television work, but you are thinking generally that television might be the career you will want to follow—with the specifics to come later. Well, perhaps this chapter can help you see the preliminaries in perspective and on a more personal level.

It is true that some jobs in television do not require a college education. That is true of many businesses in the United States. It is also quite certain that the trend in TV, as in other fields, will continue toward expectations that all newly hired employees be graduates of accredited institutions of higher learning. It may still be years to the time when TV employees are required to earn doctorates or master's degrees—but that's another story.

If you are in high school, my advice is simple and direct and with no extenuating circumstances. Bluntly put: finish high school, graduate, receive your diploma. No if's, and's, or but's. If you are in your first year at college, you ought to consider—providing you wish to pursue an interest in television—completing the required four years for your sheepskin. Statistics show that most broadcast employees who rise to middle-management or management positions in television are college graduates. The better jobs go to college graduates. The technicians who are college graduates are the ones whose status improves. Progress and opportunity appear to develop and increase almost in direct proportion to the extent of formal education. Thus, you could possibly be in sales with only a high-school diploma, but you'll progress further and faster in sales if meanwhile you are taking college courses, and you'll open the door wider for a job in middle or upper management if you also produce a college degree in addition to substantial sales.

Minimum technical training may be sufficient to operate a boom, a camera, a studio console. But as we explained in some detail in the chapters on engineering, you will better understand the equipment and how or why it should be acquired by a station plus the factors of costs for broadcast management, if you have college credentials. Business administration, as the earlier chapters pointed out, is becoming essential for the broadcast engineer who would advance in his field. There is no limit to the sophistication, acquired through education and experience, that in the future may be demanded of you in the broadcast field.

Competition in television has become razor sharp. Without a college degree in your background, you could jeopardize your future. Though it's quite possible to "train on the job," it's most difficult to be educated *academically* while working full-time in the field.

Types of Education

There's a contradiction in the broadcast industry in the viewpoints of broadcasters toward those who would enter it. Industry people tell students that there are many broadcast schools now available, and that in addition there are colleges that offer special broadcast courses. Other industry spokesmen, at the networks and in personnel, may differ. They scoff at broadcast degrees and at young people who specialize in a two-year broadcast-oriented program at a junior college or community college. Generally, broadcast executives are not plugged in to this training, nor do they believe in it. They advocate more extensive, formal education. Quite often, too, these are the very executives who will be interviewing you.

It is always fortunate when a young person has direct contact with some person in the broadcast field. The best, most direct way to break into the industry is to know somebody in it who can at least help channel you to the right person or persons with whom you can apply for a position or at least, be counseled. The next best way is to target your effort toward the department in which you wish to work. If it is sales, see if you can meet the sales manager. The manager may not be the person who ultimately will hire you. But the manager can help you the most once he or she realizes that your interest is in the domain over which the manager reigns. Would it not be the chief engineer you would rather talk with if you wanted some advice on becoming a cameraman? Or if that's shooting too high, maybe you could get to speak to a director, a studio supervisor, or the like.

Of course it's not always easy to talk to somebody in the field. It may be that you will not gain access, irrespective of what ingenious methods you may devise in getting to talk to a "live one." So in any event, whether or not you are meeting with a person employed in broadcasting or an allied field (I strongly recommend film houses that do work for television, advertising agencies that buy time, newspapers if there are joint ownerships with the broadcast station), you should also be considering a letter-writing campaign.

Job-Hunting

Write to the station or network for which you would like to work. Include information on your educational background and work experience. You must include all extracurricular activity. Drama courses, debating team, school newspaper (writing for it or selling advertising), campus broadcast station work, and club activities are all excellent areas to discuss. These should be spelled out on paper. Basic facts are sufficient. Don't overlard or overload. Time is money to your employer, and he hardly wishes to waste it by reading your memoirs.

It is highly likely that as an employee new to broadcasting your career will start at a small television (or perhaps radio) station. As a rule, experience requirements at such an outlet will not be so demanding, and they will permit opportunity for you to exploit your strong points. The big advantage in such a place, however, is the opportunity to work firsthand at many different jobs in several departments. Salaries, unfortunately, may be appreciably lower at the smaller station.

With experience gained at a smaller station, you should be in a good position, after a reasonable time interval, to advance to a larger station in a larger city. Of course, this path may not appeal to you, and you may remain at the station, as others have done, to continue a normal, rewarding, and interesting career.

The Résumé

You will have improved your chances to get the job you seek in television if you have prepared yourself "on paper." As we have seen, education is necessary, and motivation is a must. But presentation is more often the hang-up.

It was customary only a few years ago to treat the résumé as though it were a data sheet, a summary of a plethora of facts and fancy. If it recited facts only, it still usually amounted to a summary of where one had been employed and in what position.

That's not quite the way to present yourself on paper, particularly for a job in television, a field that places high value on the ability to communicate. For an entrée, then, we recommend that you communicate!

You must communicate *you*. Often in job-hunting in television—in fact as well as in story—professional talent (announcer, disc jockey, news personality) communicates by voice (often on tape) or by image (photographs, film, videotape). But as a practical approach for the average gal or guy, the *résumé,* the *interview,* and on occasion the *tryout* are the "openers" in television job-seeking.

So let's take on the résumé, with advice of the experts who've studied it and of the industry's pundits who have their interpretations.

One thing you'll come up with after these consultations is agreement that an applicant's recitation of the dull facts, irrespective of the authenticity or documentation, isn't sufficient to fulfill requirements established by today's sophistication.

Preparing for Careers 97

There are several points to consider. (We can conduct this as an exercise in order to emphasize our various points.)

Take a sheet of paper and jot down the numbers one through four, leaving space beside each number for notations. Next to #1, write "Selling a service." We'll use that phrase because this is really what you will be doing. You will be *selling a service*. It's up to you to tell your future employer what you can do, what service you think you can sell to him. Good grades in English, for example, are fine but they are expected.

In effect, you should be writing up a sales sheet for a résumé. It's a rule among today's résumé writers to determine what job it is they really want and then to design the résumé to fit those requirements—a general litany of jobs and schooling may not go far enough.

As #2 on your list, write "Don't read between the lines." It is better to write a résumé for a particular job. In this résumé, or sales sheet, one should write about job functions performed or about which one has a particular knowledge or skill. This is in contrast to a recital of the number of years spent at old Gaylord U., although this information is important and must be included.

You should tailor the information for a specific function; don't leave it up to your would-be employer to guess for what job you would best be suited.

Next to #3 write "Functional résumé." And at #4, write "Orderly and brief." (The *functional* résumé is to be distinguished from the *chronological* résumé.)

Writing the Résumé

Aside from the actual text, the résumé should contain your name, address, and telephone number, all three usually at the top left of the page; and at the right, the title of the job you seek.

Topically, your experience is the first area you'll write about. Don't hedge, but go into experience immediately. Cite summer work you've done; part-time jobs—all work, including nonpaid volunteer positions. Show yourself to be a doer (if you are one, and if you are interested in TV as a career, it'll be more comfortable for you if you *are* a doer). Provide examples of work, what problems you encountered, and the methods used to solve those problems.

Next move into skills. What are they? Do you type, for example? For television, of course, you'll want to list participation in drama clubs, plays, productions, and the like. Do you have language ability?

Now come the categories that do not fit anywhere else. Here you should include organizations with which you've been active; books or articles you've written or edited (including high-school paper, journal, yearbook, etc.); and travel (though you should state the purpose to show that it was relative).

From this point on, the résumé should be traditional in form by giving the prospective employer a full rundown (the chronological list). A report is presented showing your employment record (include your title, name of employer, and period you were there). Among the variations in putting this together, the name of the employer may be given first, with the other information following. Dates need not include the month of the year unless it has unusual pertinence to the job being applied for.

After a section on employment, data on education is next. Again this is all "necessary" information, including the schools or colleges attended. If you have a college background, it is superfluous to include high school. Give the highest level. An exception is a graduate-school degree or attendance at a graduate school; this should be added to information on undergraduate honors and degrees, or attendance. Simply list the college, degree earned (abbreviations are all right), and the year.

At the end of the résumé, personal items can be added, but don't get personal unless the information in question is needed to help you get the job. This should always be the criterion. It's really your objective in putting together a résumé in the first place.

General Aspects

Here, then, in summary are some do's and some don'ts.

Don't mention salary. Leave that open for you and the employer to discuss.

Do write one-page résumés. That makes it easy for the employer to scan your credentials.

Don't go more than two pages. The law of diminishing returns takes over after page two.

Do remember that a résumé ideally should present a picture of you, of how you could perform.

As explained elsewhere in this book, some people advise the applicant to lean heavily on any and all job experience in seeking a spot in the TV field (again, too, if you have entrée through a friend in the business who can help arrange an interview, so much the better, but be prepared to submit a résumé in any event). Some in the field suggest applying in the spring (as in promotion) or in the winter (as in engineering) for summer jobs in television and as a direct way toward applying eventually for full employment.

CHAPTER XVII

Taking It from the Top

Television, as you've been told, is growing and full of opportunity. But there is no easy access to its employment rosters. Before you can be considered employable, you will need some basic skills, specific interests, and intense motivation. Most of all, you must obtain the necessary education or training where that is a prerequisite.

The field is a big one. There will be openings for jobs and opportunity for advancement. But as volatile a field as television may be, it is also sharply competitive. Remuneration is excellent. Basic salaries, fringe benefits, and prospects for improvement are all good. Usually they are on the level of other, similar industries. In reality, the actual scales are somewhat higher in television.

A TV career offers a number of specific occupations. Some areas of endeavor also exist, such as programming, news, management, engineering, and sales—to name the key categories.

Programming sits next to glamour. It is television's honey pot, the magnet that attracts, the sizzle and the steak. Programming is what TV is all about. If that is your attraction, then your preparation must be specially skewed (the TV word for slanted).

First you must expect to earn money. It will be good money at the station level, a good hunk of change at the major station, and even more as you progress into the networks and the rarefied air of associated program companies and studios.

Glibness is a thing of the past. But it will help to have a sense of the theater and a better than working knowledge of the movies and of stars and personalities; also to be an avid reader, particularly of plays, novels, short stories, and even newspapers and magazines. General educational background is preferable, with no heavy emphasis on a specific area necessary.

Although stations may not program too much of their own schedule, the programming function is critical to the success and health of the operation.

If you are in a position to work at a campus television (or even radio) station, use that opportunity to try your hand at various departments, emphasizing programming if that is where you think you want to be.

The news area is technically a unit within the programming department at many TV stations. Of course, it is a separate function at the very large stations and networks.

Television news today is one of the more active, vibrant, and exciting aspects of the industry. It draws on the skill, work, and talent of a number of people who as a team gather, write, produce, edit, and broadcast news and public-affairs programming. News coverage has become more sophisticated, faster, and more meaningful as well as mobile with the introduction of miniaturized cameras and other equipment.

News is said, moreover, to be the fastest-growing part of the television industry. As a journalist, this is the place to be, particularly when looking toward the future. Electronic journalism is often where the action is.

There is no pat route to take in vying for a position in television news. Whether one first takes up news reporting with a newspaper or joins TV news directly from college is really immaterial. The objective should be first to get hired at a station or a network in the news department. No menial job should be too low for the asking and the taking. There's always time and room to advance.

Don't count on an opportunity to get on the air as an on-air reporter or anchorman or commentator. That is, don't count on it unless you have visual charisma—unless you look well on camera and have a pleasant voice with a good dose of authority in your

CBS PHOTO

Newly designed television camera and sound systems, portable and easily handled by the professional newsman and news technician, have become increasingly popular. They were much in evidence at the national political conventions in 1976, where they permitted television news reporters to roam almost at will on the convention floor for interviews and commentary.

manner. If you cannot hack it in front of the camera, wipe any such aspirations from your mind and concentrate on a support role. There are many of them at every station.

Education and some experience are extremely important in television news. Your education should center on a big appetite for politics, community events, current events, and just plain ordinary people —all kinds of people. You should enjoy talking and dealing with people. Preferably you will have a college degree, be well read, and have majored (or at least minored) in English and literature. The best advice is to start as a writer at a local station. It's a sound basis, and it's likely that a writer will be considered for reporting in the field.

Bright young business-trained or liberal-arts-oriented people who understand business, particularly the buying and selling processes, are the element most often considered for station management positions. A background in sales, in particular, seems to be the usual for most managers in television operations around the country. A station manager (or program manager or general sales manager or chief engineer) must be literate, intelligent, and articulate. Ideally he is wedded to, and committed to, the community and its problems.

The college course laid out for the management aspirant is liberal arts, with an emphasis on courses in economics, marketing, and other facets of business administration. Sales is the way to learn about the broadcast business "on the job." It is also the traditional training ground for future managers.

Engineering in the broadcast business is undergoing change. Whereas once technical training was enough—perhaps even an apprenticeship would have sufficed—it is now necessary to obtain a formal education. This is so if you aspire above the level of technician. If your goal is even higher than middle management, it might be advisable to consider an electrical engineering degree topped with a master's degree in business administration.

New electronic equipment is becoming an everyday occurrence in the broadcast industry. Many consider the technical area of the relatively young TV industry to be the most promising and challenging. Certainly if technical work in television appeals to you, the field can offer very stimulating situations. And, most likely, a small station will offer you the most at the beginning of your career. At a small local station you will be able to handle a varied assemblage of technical chores. You will learn faster and more.

Engineering in television has also become part of a sophisticated business in which satellites and space technology are naturally involved and integrated. Simply stated, the engineer's job is to make it technically possible to get the program out of the station's studios and to the transmitter and into the home. But with that function has grown a complex, giant, and intricately engineered communications system that employs thousands of technicians, nearly all expert in their fields.

That the people of the United States have grown accustomed to expertise in TV engineering was brought home dramatically and with no little embarrassment to the American Broadcasting Company when the sound transmission was lost during the Carter-Ford Presidential debate held in Philadelphia in 1976. The TV industry figuratively shuddered, with all competing networks and TV stations, affiliated with ABC or with the other networks, sharing in the horror of an industry that had followed the giant steps of mankind on the moon tripping over an easy chore like the transmission of sound from a metropolitan city.

But the jobs entailed in TV engineering are essential to the industry. Performers may have days off, sabbaticals, and "off days," but television as a service is on, presumably at its best, every minute and hour of the day and night that the station is transmitting its signal.

Many of the changes in programming, especially of live events as in news happenings and sports activities, have been ushered into the 1970's by the changes in equipment and engineering. TV is now mobile, and it can replay, freeze action, and slow the action. It can follow a President to the Great Wall of China, and its cameras can be hand-held, walking alongside the President of the United States as he takes a stroll on the White House grounds.

Television broadcast has reacted to the new technology by adapting to its ways and adopting its most glamorous and effective methods. At the same time, the decline in demand for engineers in the country has had its effect on the television industry. The abundance of trained, highly qualified technicians and scientists has served to stress the stability of engineering employment and also the desirability of getting into the field.

The rise of the design engineer in the TV engineering picture is new. This person has assumed importance along with the heightened development and use of the new technology. Yet he is clued to the TV industry because the design engineer must also serve as a businessman. Industry leaders recommend that the design engineer (with degree) pursue also the study of business administration, usually at the graduate-school level.

Another name for the new technology is "computer technology." The computer design in electronics has come of age.

While the programmer produces the product for the industry, management keeps everything smooth and efficient and sailing, and engineering gets the product out of the studio or out of the film can, it is sales that keeps the whole commercial TV system alive. Broadcasters are dedicated to service to the public. They promise entertainment and information, amusement and diversion as well as enlightenment and even education. But they do not operate a charitable institution. They operate a business. They sell time.

The immediate or instant springboard to an insight and knowledge of commercial TV operations rests with the sales function. Selling time for a TV station or network brings the person into contact with advertisers and their agencies on the outside; the programmers, business operations people, and skilled management on the inside. The interplay among all these interests amounts to a learning experience for the salesperson. It's a sure way to a highly pressurized executive training, on the job and, most likely, on the firing line.

Competent salespeople usually are welcomed in the television business. Youth has an advantage and is favored in this frame of reference. Yet it may not be just a hop, skip, and jump into a sales position. Experience in selling on a school paper or campus station may be just what the doctor ordered. It can help you land the job, and it's quite possible that you may have to settle for less somewhere else, at least at the outset of your career in sales. Once you accumulate experience in selling, a TV station or network may be more inclined to consider your credentials.

Sandwiched somewhere between programming, management, and sales at a typical major TV station operation is the promotion department. This office may range from a one-man or one-woman department to many people with department heads, operational managers, and subdepartments of advertising people, public relations people, and photographers. Perhaps the visual epitome of TV flexibility, uncertainty, feverish activity, and talented imagination is contained in the promotion arm. Yet in many parts of the country, this area is still unsung, underpaid, overworked, misunderstood, misjudged and "the first to go" in times of cutbacks and economizing.

Promotion people may be sensitive, well-educated, well-rounded, able individuals. But they are also probably the most dependent department of any at the typical station or network. Broadcasters are convinced that they cannot do without programmers, salespeople, or engineers, but that they can do without a promotion man or woman indefinitely. That calls for a finesse among promotional people that demands the most subtle minds and emotions, because if they are permitted to assert themselves they can make enormous contributions to a station in terms of image within a community, of audience growth, of bigger profits subsequently, and of morale. The latter—morale—should not be overlooked. It is important in various industries of diverse interests. It is especially so to a communications company. What people working at the company think of themselves and of the company will inevitably be projected to the people they deal with and finally to the public itself. Closely associated with this area then (i.e., public relations) is community relations. Only the largest stations and the networks have people so involved. But today, with stations undergoing pressures from regulatory agencies in Washington, from minority groups and the disadvantaged, the role of community relations expert in the TV industry ranks high with the company's executives.

In the promotion orbit, formal educational background looms as important, if not essential. A promotion applicant should have writing talent, a command of English (ease in sentence structure, syntax, and vocabulary), and be a student of liberal arts, in undergraduate school but eyeing continuation on the graduate-school level. Editorial work on a school paper would he helpful. Exposure to the news syndrome can catapult a young person from the rock-bottom rung of the ladder in promotion to a position almost anywhere on up. Some industry spokesmen suggest that hopefuls use news training as an acquired skill that would be exercised whenever the applicants are called upon by station management.

Potentially, unions will be the organizations with which you'll have the most contact other than with your fellow employees. It may not be so important whether or not you like the union, but it could be vital to your interest to know what a union *is*, how and/or why it wants to help you as an individual. The advice is to get along with unions. Know what makes them tick and how they can make life easier or perhaps more difficult and trying.

The nucleus, to an extent, of unionization in the television industry is made up of the three television networks and their owned and operated stations and extensive facilities in such metropolitan areas as New York, Chicago, and Los Angeles. Add to this a large number of major TV stations throughout the land, and the power and size of the unions in the industry become clear.

The television and radio stations in the large metropolitan areas tend to be more heavily unionized than those in the smaller communities. Technicians, such as the cameramen, stagehands, lighting engineer, and sound man, are usually heavily organized. Their unions are tightly drawn. Announcers and writers, however, are likely to have more latitude in their unions, and salespeople and programmers would not be expected to join unions.

Television centers in two of the largest cities in the country: New York and Los Angeles. Both are union cities. New York is historically strong union. Los Angeles is, too, mainly through Hollywood, where the film studios have been the hub of union activity.

It is suggested that the smaller station is loosely organized, if it is organized at all. And, while we do not challenge the union's position in broadcasting, we do suggest that for some young people entering the television field the smaller or nonorganized station might be the preferable place for them to start their careers.

There is the concept of unionization. By joining a union, voluntarily or involuntarily, you give up some of your freedom of individuality for some additional security.

Although the television industry has made great strides in its employment practices, it perhaps is moving fastest in keeping pace with employment trends in relation to the hiring of minorities and women. At the time of writing this book, the National Broadcasting Company was reported to have agreed to a $2 million out-of-court settlement of a sex bias suit brought by women employes.

The most interesting aspect of this settlement was that the terms included not only provisions for back pay but promises also of better job assignments and of increased hiring of women. NBC was said in the agreement to provide equalization money for some of the 2,800 women (who had worked, now worked, and expected to work) in certain jobs. This part of the agreement was designed to give women equal pay levels with men in similar jobs, thus admitting that women at NBC had not been promoted, or had been promoted at slower rates than men. Other provisions included goals for the hiring of professional and managerial women and promises to improve the woman's lot in technical jobs.

Broadcasting, nevertheless, is open—perhaps more than other fields—to the employment and advancement of women. Most people in the industry believe there is a better climate for women in broadcasting than in other businesses. In addition, there is a constant effort to upgrade women's role in the television industry. Pay in some areas, as in news, is about equal for men and women. Conditions in the broadcast industry for the most part are naturally encouraging as well as tolerable for women.

While women are achieving a near equality with men for employment opportunities and salary considerations in broadcasting, there are still roadblocks for many women to get through before they can expect to gain an equal, solid footing in the industry.

In the mid-1970's the opportunities for jobs in the broadcast industry are no better and no worse than they may be in other businesses. But at no time in history has the opportunity been as good in television as it is now for women to start their careers.

According to the U.S. government, the percentage of minority-group in broadcasting increased over a six-year span from 9.4 percent, or 6,572 persons out of a total of 70,056 employed in 1970, to 15 percent, or 11,867 persons out of a total 79,295 employed in

1975. The Equal Employment Opportunity Commission published these and other figures on minority groups, which it classified as blacks, Asian Americans, Spanish-speaking Americans, and American Indians. Aside from these figures, the commission found that the total number of women in broadcasting had increased from 17,773 in 1970 to 22,258 in 1975 (and minority women went from 1,925 to 4,906). A survey by The United Church of Christ, meanwhile, reported that in 1976 the proportion of minority and women workers in television-station employment increased 1 percent for each group. In that year, minorities held 14 percent of the jobs, or 5,769 positions, and women held 26 percent, or 10,871 positions.

In previous chapters, some mention was made of one or more aspects of the television industry. One whole chapter, however, was devoted to the dimensions of television. In that chapter, we found that television is not only a growth industry, but already a giant business. Sometimes its full dimensions in the economy and in the life-styles of the country are not realized sufficiently by those engaged in the daily routine of the business.

The commercial impact of television is awesome. Revenues in one year alone in the 1970's passed $4 billion, while profits soared over $780 million.

In plotting television's dimensions, one must include in them a measure of the television networks. In any study of networks and the relationship to employment, it is clear that the operation requires experience before advancement can be offered, and that before there is development of a newly hired individual there must be talent and determination. One might add to this—motivation.

Among the corporate creatures spawned by the dynamics of television are the station groups. The three television networks are themselves the parent companies of subsidiaries that maintain ownership and operation of television stations. These subsidiaries are station-group owners. In addition, there are numbers of other independent station groups or groups owned in turn by such giant corporations as Newhouse, Westinghouse, General Electric, Kaiser, Cox, Wometco, and Post-Newsweek. One of the networks, NBC, is the parent of a group of stations, and NBC in turn is owned by RCA.

While the assessment of commercial television is that this form has fostered creativity, competitiveness, and bold, bright entertainment, there is obvious room for other services, such as noncommercial television, or as it is commonly called, public television.

Today's public TV is itself a complex of networks. But it is also almost a contradiction: on the one hand is the pull of a desire to program as a network and control entertainment on individual noncommercial stations nationally; and on the other hand is a commitment to program locally, station by station, with the individual, community-minded facility deciding what to program and how and refusing a network feed if it so wishes.

Be that as it may, public television is growing at a comfortable pace, with funding available from various sources, including federal allocations. The appropriations come principally from Congress, though other funding has come from such agencies as the Department of Health, Education, and Welfare.

Despite all this activity, public TV at best is only another area offering additional opportunity for those who would choose a career in television. Constructively, public TV has traditionally displayed a need for young people with stimulating ideas, determination, and dedication. The critical jobs would seem to be those of general, program, and production managers.

Aside from public television, the medium embraces an expanding industry that already includes the viable cable television, pay television, and closed-circuit television. An explanation of each would give you repetition rather than provide a summary. But it may be worth emphasizing a point.

For example, as a young person considering a career in the relatively young and somewhat fledgling field, you must keep your eyes open and your attention diligent to any nuances denoting change or expansion. So that while the noncommercial aspects of TV may appear remote or unattractive, don't sweep them aside in a pile of discards. Keep them in mind for future reference. They may come in handy. They may be your salvation in a sense.

No matter which route you take in television or what aspect may appeal to you the most, the field may be as closed to you as a flower during a cold spring night. It may very well stay closed unless you can warm things up and gain reception. Your attitudes must express warmth, understanding, enthusiasm, an appetite for change, a zest for getting things done, a sense of public relations, creativity, a balanced temperament, reliability, and initiative. All of them? Well, a good amount of these attributes would be helpful in keeping you not only happy and busy, but needed, wanted, appreciated, and rewarded.

What we've done in this chapter is literally to take it from the top in the sense that we have summarized the message. Sometime ago Marshall McLuhan coined the phrase, "The medium is the message" when writing of television. Perhaps what we are saying here is that the message is really the medium. What we wish to convey is that television is unique because there is truly nothing like it. The closest one can come is motion pictures. But if even that is true, where does this leave radio? The newspaper? The stage? Are they not all either copied or competed with by television? And does not TV content contain a little of each one's format? TV sells time like radio; has announcers like radio; operates with commercial cut-ins and schedules like radio, etc. TV summarizes the top news as do most of the large newspapers (although they also go into individual stories deeply). TV features news and other events almost like magazines. Television has big stars who dance, sing, and tread the stage boards as on Broadway. TV does so many things like so many others.

Although repetition is to be avoided, it is necessary to mention that a college education is becoming a necessary item in preparing for a career in television. It is assuming an even greater importance in the 1970's than only ten years back. By the 1980's the education prerequisites for employment in the television industry can be expected to have become more formidable. The previous discussions of engineering in the broadcast industry should have underscored that point.

A point worth remembering is that you may wish to start your career in television at a smaller station and possibly also in a smaller city. That way you may learn more about the business in a relatively shorter period of time than if you tackled the powerful station in the big city. And the smaller entity may have more reason to hire you, if only because it receives proportionately fewer applications.

In any event, learn how to work up your application. Invariably the application should be submitted with a résumé, unless you are instructed to do otherwise. (Don't be bashful; ask the station or network if you haven't been advised.)

Television is a world of its own. It has many areas open to talented, gifted, or—most of all—enthusiastic young men and women. Lawyers, certified public accountants, clerks, bookkeepers, as well as actors, announcers, performers, cameramen, producers, directors, management people, sales executives, engineers, and newsmen are all needed in, of course, different capacities and at various times. Because of the diversity in many of the careers (and in fact, they often overlap), television remains a vigorous, volatile field in which many opportunities present themselves. There is constant movement within the industry. At times, this movement is like a rush of water. Once you try it and find the temperature to your liking, you, too, may decide to dive in and head for the open. Keep your head up, your arms strong and grasping, and enjoy the swim—the water, you know, is fine.

Bibliography

Broadcasting Publications. *Broadcasting Cable Sourcebook.* Washington, D.C., 1977.
———. *Broadcasting Magazine.* Washington, D.C.
———. *Broadcasting Yearbook.* Annual. Washington, D.C.
Deutscher, Noel J. *Your Future in Television.* Rev. ed. New York: Richards Rosen Press, Inc., 1963.
"Facts, Figures & Film, News for Television Executives." Special NATPE Conference Issue, February, 1977. New York: Broadcast Information Bureau.
Hilliard, Robert L., ed. *Understanding Television.* New York: Hastings House, 1964.
Jencks, Richard W. Television News and "The Golden Age." Address, CBS Television Network Affiliates. Los Angeles: CBS, Inc., 1970.
Job Opportunities, Description of. New York: American Broadcasting Companies, Inc.
Lawton, Sherman Paxton. *The Modern Broadcaster: The Station Book.* New York: Harper, 1961.
Mayer, Martin. *About Television.* New York: Harper & Row, 1962.
National Association of Broadcasters. "Careers in Radio." Washington, D.C., 1974.
———. "Careers in Television." Washington, D.C., 1974.
New York Times. "Hiring of Minorities Shows an Increase," January 3, 1977, p. 43; "Women Minorities Raised Proportion of TV Jobs by 1%," January 10, 1977, p. 44; "$2-Million Settlement Is Reported in Women's Suit Against NBC," February 13, 1977, p. 1;

"$2-Million NBC Pact Is Set as a Settlement with Women of Staff," February 17, 1977.

Nielsen Television 1977. Northbrook, Ill.: A. C. Nielsen Co., 1977.

Quaal, Ward L., and Martin, Leo A. *Broadcast Management: Radio, Television.* New York: Hastings House, 1968.

Roe, Yale, ed. *Television Station Management.* New York: Hastings House, 1964.

U.S. Department of Labor, Bureau of Labor Statistics. "Radio and Television Broadcasting Occupations," in *Occupational Outlook Handbook.* Washington, D.C.: U.S. Government Printing Office, 1972–73 and subsequent eds.